T0353507

Fundamental Aspects of Corrosion Films in Corrosion Science

Fundamental Aspects of Corrosion Films in Corrosion Science

Bruce D. Craig
Metallurgical Consultants, Inc.
Houston, Texas

Plenum Press • New York and London

Library of Congress Cataloging-in-Publication Data

Craig, Bruce D.
Fundamental aspects of corrosion films in corrosion science /
Bruce D. Craig.
p. cm.
Includes bibliographical references and index.
ISBN 0-306-43623-X
1. Corrosion and anti-corrosives. I. Title.
TA418.74.C73 1990
620.1'1223--dc20 90-48814
CIP

ISBN 0-306-43623-X

A Division of Plenum Publishing Corporation
233 Spring Street, New York, N.Y. 10013

Printed in the United States of America

This book is dedicated to my sons,
Jason and Nathan.

Preface

The purpose of this book is to stimulate thinking among corrosion scientists and engineers to examine corrosion mechanisms and corrosion control from another perspective. While the presence of corrosion films in electrochemical corrosion has been recognized for over a century, the contribution of these films to all facets of corrosion has not been explored to a significant degree. Rather the role of films in certain mechanisms (i.e., stress corrosion cracking) has been emphasized, yet almost ignored for other corrosion mechanisms. This is viewed by the author as solely attributable to the lack of investigation into, and an understanding of, the contribution of films to these mechanisms or forms of attack.

The lack of emphasis and study of corrosion films and their contribution to all forms of corrosion attack are probably the result of current university instruction that utilizes two popular corrosion texts (Uhlig and Fontana and Greene) for teaching. These texts provide an excellent understanding at the undergraduate level of corrosion fundamentals; however, the major implicit premise in these texts is that bulk properties of an alloy or metal control the corrosion resistance in a particular environment. For many applications and for a simple understanding of corrosion mechanics, this approach is sufficient. Yet, research on corrosion films indicate these films often have an entirely different composition than the bulk metal (ratio of alloying elements). Moreover, the film may be protective in certain environments and nonprotective in others. Furthermore, once a film is established, transport through the film and reactions at the metal/film and film/environment interfaces are in many cases rate controlling as well as mechanistically controlling. Therefore, since most metallic systems rapidly form some type of film when exposed to an environment, the environment/film/metal system becomes more important in evaluating alloy–environment behavior than just corrosion behavior of the bulk alloy.

This book is not intended to be a theoretical text on oxide films, since this has been dealt with elsewhere; instead it is meant to complement the more basic texts on corrosion, allowing the student/scientist to extend this basic knowledge of corrosion fundamentals to an appreciation of the role of films in corrosion. As such it completely emphasizes films, and no attempt has been made to address other factors that at times may be of equal or greater importance. It is further hoped that corrosion scientists, engineers, and students will gain a new perspective on the importance of corrosion films in corrosion science and include this insight in further work.

The obvious role of corrosion films in the future will be complete control of corrosion by manipulation of the film. In other words, it is hoped that future research will provide sufficient insight into all aspects of corrosion films so that the surface layers of a metal can be alloyed to grow the necessary film to create protection and provide a self-healing system when the film is damaged in service. While this eventual goal is certainly many years in the future, current research suggests that such goals are generally achievable. Thus, engineering the bulk or surface alloy content of a metal to produce a protective film for a specific service environment is the future emphasis of corrosion films, and it is the purpose of this book to be a guidepost to this goal.

The book is divided into seven chapters. Chapters 1 and 2, review thermodynamics and kinetics, with emphasis on corrosion films. Chapter 3 represents the main thrust of the book and is intended to provide the reader with a fundamental understanding of films and how they affect corrosion mechanisms. This chapter is an introduction to transport in films and some of the important models currently used to describe films. As with Chapters 1 and 2, there are other texts that are more theoretically detailed and should be consulted if a deeper knowledge of this topic is desired. Chapters 4–7 specifically address the roles of films in the areas of corrosion inhibition, pitting, environmental cracking, and erosion-corrosion.

While films contribute to essentially every form of corrosion, these certain forms of corrosion were chosen because of their importance to corrosion research.

Acknowledgments

It is extremely difficult to write a book on a subject of this magnitude without failing to acknowledge the considerable contribution of the many researchers in this field. Therefore, I express gratitude to all those people referenced in this book and especially to those not mentioned that have advanced the science of corrosion.

I would personally like to thank Professor David L. Olson of the Colorado School of Mines: first, for his enthusiasm in teaching the subject of this book, which inspired my interest, and second, as a friend who has championed my cause from the beginning of my career. No one can ask for a better example of a teacher. Furthermore, I would like to acknowledge Professors William Mueller and William Copeland for their support in achieving my academic goals. Without their early assistance this book would not have been possible. Appreciation is also extended to Dr. Donald T. Klodt, who stepped in to assist after my first technical paper was rejected and, thereafter, encouraged me to continue publishing.

Finally, I thank two important women for their support and assistance in the substantial effort of this book: my secretary, Dannia Lou Stovall, and my wife, Dolly.

Contents

1

Thermodynamics of Corrosion

Introduction

The thermodynamic aspects of corrosion and, more generally, electrochemistry are quite complex, yet are most important to understanding the driving forces behind them. This chapter sets the stage for the remaining chapters on surface films; it is not a detailed dissertation on the thermodynamics of corrosion. Numerous excellent texts are available on the concepts and theories of electrochemistry and thermodynamics, yet few focus specifically on the electrochemical aspects of corrosion.

This chapter provides the foundations for a deeper understanding of the driving force for the reaction of aqueous solutions and gases at high temperature with metallic surfaces.

Charged Surfaces

Almost every surface or interface, regardless of the material considered, has a net electric charge. Consider, for example, metals; the metallic bond is characterized by a sharing of valence electrons between numerous metallic nuclei. At the surface, in the presence of a vacuum, valence electrons will remain mobile but are no longer shared beyond the surface, resulting, on average, in a net negative charge adjacent to the surface. Nearest the surface within the metal a net positive charge will arise due to the position of the nuclei and the necessity for charge neutrality in the near interface region. However, the interface as viewed from outside the metal has a net negative charge. This is shown schematically and simplistically in Fig. 1.

Of course, once the metal is introduced to any environment other than a vacuum, chemical species in that environment will interact with the charged surface. The study of this interaction is the broad area referred to as *surface science,* and at this junction this book could diverge into numerous focused treatises. However, it is the purpose here to emphasize the role of surfaces as they relate to the interaction with their environment, resulting in the production of surface films.

The next step is to follow the behavior of this negatively charged surface as it is exposed to an active environment, such as air or water vapor. The charge concentration at the surface does not actually exist as an abrupt singular interface but extends as an electric field a very short distance into the environment. In the presence of water molecules, which are asymmetric in structure, the positive end of the water dipole will be attracted to the negative electric field generated by the metal surface. This results in the negative portion of the water dipole facing out into the environment. Similarly, positive cations of other chemical species will be attracted to the charged interface. Since

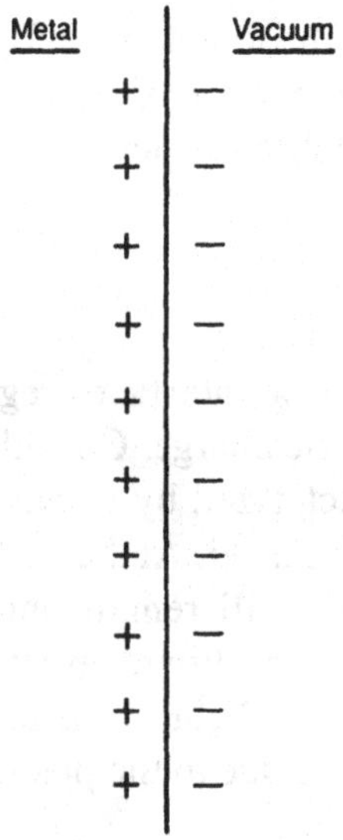

Figure 1. The charged interface of a metal surface in vacuum.

cations have a strong electric field of their own, they have a tendency to accumulate a sheath of water with the water dipoles oriented in the same manner as for the charged interface, with the positive pole directed away from the cation. Thus, cations approach a charged interface surrounded by a sheath of water molecules oriented with a net positive charge outside the sheath. Since the charged surface is already covered by a monolayer of water molecules the solvated (sheathed) cations are limited to a specific distance from the charged interface, referred to as the *outer Helmholtz plane* (OHP) (Fig. 2).

While electrostatic considerations have a profound effect on the attraction of ions to the metal surface, chemical gradients can result in the adsorption of anions simply by contact at the charged interface. Thus a combination of anions and water will occupy the monolayer adjacent to the charged interface, which is referred to as the *inner Helmholtz plane* (IHP) [1].

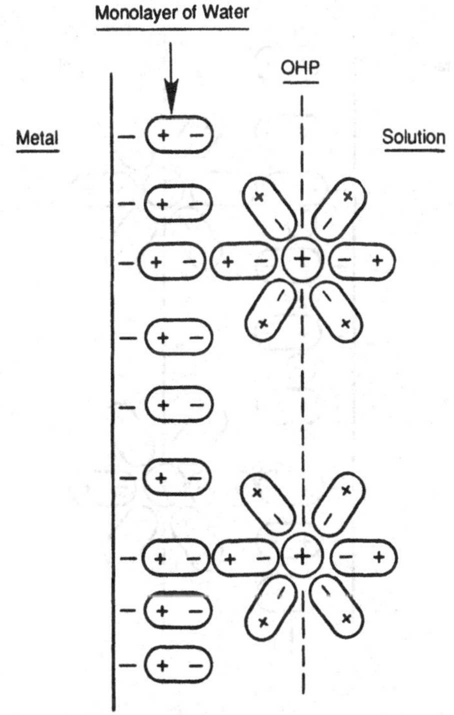

Figure 2. Solvated cations restricted to the outer Helmholtz plan (OHP) by the dipole nature of water molecules.

Figure 3 illustrates the general concept of the charged interface with the IHP and OHP [1]. The metal surface will invariably produce an excess concentration of electrons compared with the number of adsorbed water molecules, thereby creating an excess charge density at the interface. In order to retain charge neutrality, a similar positive charge produced by cations will accumulate at the OHP. These two layers of charge are often termed the *double layer* and are of considerable importance, both to the electrochemistry of corrosion and to the formation and behavior of corrosion films. Conceptually, the double layer is analogous to a capacitor, with one plate equivalent to the excess charge of the metal surface and the other to the balancing charge of cations in the OHP [1] (Fig. 4). The advantage of using

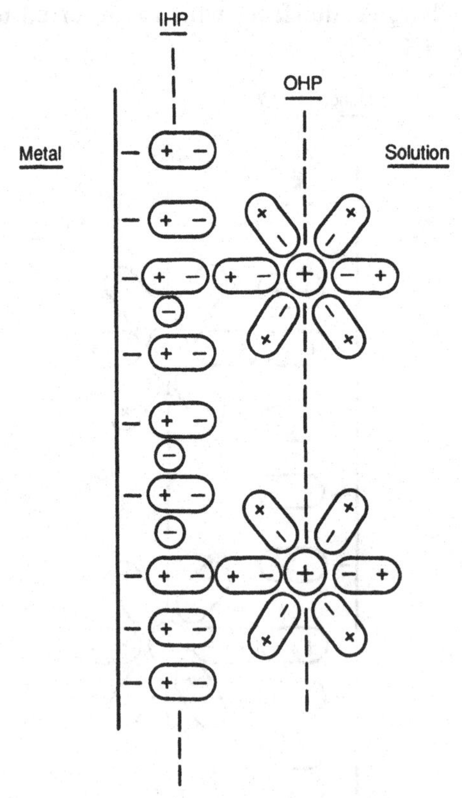

Figure 3. Location of the inner Helmholtz plane (IHP) and OHP with their associated ions.

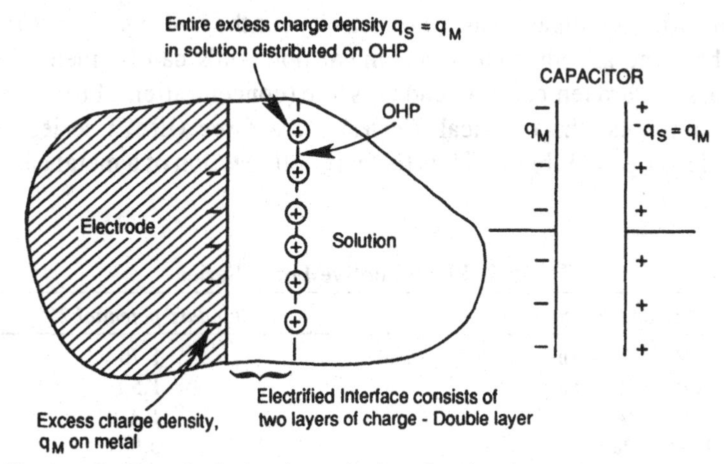

Figure 4. Electrical circuit equivalent for the charged double layer.

electrical circuits to illustrate certain aspects of electrochemical behavior will be emphasized in later chapters.

Of course, the presence of these excess charges will produce a corresponding potential that is negative at the charged metal interface and becomes more positive as the OHP is approached. This will create an electric field across the double layer that is characteristic of the specific metal/solution combination. The potential at these interfaces cannot be directly measured, but the potential of this entire electrode can be measured against reference or standard electrode.

Electrochemical Reactions

Mineral beneficiation and extractive metallurgy are the art and science of producing a metal from its ore (refining). These processes require the input of energy which becomes stored energy and is available to supply the driving force (free energy) for return of the metal to its native (oxide or sulfide) state. This process of degradation of a metal by the environment is called *corrosion*. The tendency to return to its native state is a measure of its free energy and can be related to any of several standards. The most common standard electrode in electrochemistry is the hydrogen electrode. The *standard hydrogen electrode* (SHE) is designated 0.0 V when the electrode is at 25°C,

1 atm hydrogen fugacity and unit activity for the hydrogen ion. From this baseline all other electrochemical potentials can be measured, also at standard temperature and pressure (concentration). From these measurements, the classical *electromotive force* (EMF) series has been developed (Table 1). Thus the half-cell reaction of a metal can be

Table 1. Electromotive Force Series

Electrode reaction		*Standard potential 25°*
$Au^{2+} + 2e = Au$	Noble	1.70
$Au^{3+} + 3e = Au$		1.50
$Pt^{2+} + 2e = Pt$		1.2
$Pd^{2+} + 2e = Pd$		0.987
$Hg^{2+} + 2e = Hg$		0.854
$Ag^{+} + e = Ag$		0.800
$Cu^{+} + e = Cu$		0.521
$Cu^{2+} + 2e = Cu$		0.337
$2H^{+} + 2e = H_2$		0.000
$Pb^{2+} + 2e = Pb$		−0.126
$Sn^{2+} + 2e = Sn$		−0.136
$Ni^{2+} + 2e = Ni$		−0.250
$Co^{2+} + 2e = Co$		−0.277
$Tl^{+} + e = Tl$		−0.336
$In^{3+} + 3e = In$		−0.342
$Cd^{2+} + 2e = Cd$		−0.403
$Fe^{2+} + 2e = Fe$		−0.440
$Ga^{3+} + 3e = Ga$		−0.53
$Cr^{3+} + 3e = Cr$		−0.74
$Zn^{2+} + 2e = Zn$		−0.763
$Cr^{2+} + 2e = Cr$		−0.91
$Mn^{2+} + 2e = Mn$		−1.18
$Zr^{4+} + 4e = Zr$		−1.53
$Ti^{2+} + 2e = Ti$		−1.63
$Al^{3+} + 3e = Al$		−1.66
$Hf^{4+} + 4e = Hf$		−1.70
$U^{3+} + 3e = U$		−1.80
$Be^{2+} + 2e = Be$		−1.85
$Mg^{2+} + 2e = Mg$		−2.37
$Na^{+} + e = Na$		−2.71
$Ca^{2+} + 2e = Ca$		−2.87
$K^{+} + e = K$		−2.93
$Li^{+} + e = Li$	Active	−3.05

related to the standard hydrogen electrode and its deviation from the standard. That is, the more negative is the deviation of the half-cell reaction from the hydrogen electrode, the greater is the driving force for corrosion. Conversely, the more positive the deviation from the hydrogen electrode, the less likely corrosion will occur. Of course, thermodynamics will not predict the *rate* of corrosion, only the direction or tendency. Furthermore, to be valid thermodynamically, all these reactions must be reversible.

When the two electrodes of a cell are connected together, a current will flow, as a result of the potential difference described by the electromotive force. The electromotive force is the driving force for the reaction at each electrode. Each corrosion cell will have a characteristic potential, and since each cell can be divided into half-cells, each of these will also have a characteristic potential.

The EMF series ranks half-cell reactions in reference to the hydrogen standard in terms of volts. These potentials are defined as *standard electrode potentials;* for which all of the reactants and products are at unit activity. However, in practice this often translates to 1 *M* concentration for ions, 1 atm pressure for gases, and solid phases are defined as unity. The standard temperature is 25°C. The EMF series in Table 1 presents reactions as reduction reactions; they may also be presented as oxidation reactions.

As mentioned, the EMF is the driving force for a reaction but so also is the Gibbs free energy; thus, the relation

$$G = -nEF \tag{1}$$

holds, whereby the electrical functions of a cell may be related to the chemical aspects. Here n is the number of electrons taking part in the reaction, F is the Faraday constant, and E is the potential of the standard half-cell reaction.

Although the standard hydrogen electrode is widely accepted for the EMF series, it is not convenient to perform electrochemical measurements using a hydrogen electrode. Therefore, many other more convenient reference electrodes have been developed and are more widely used in corrosion research. Two of the most common are the silver/silver chloride and the calomel. Since most investigations are only interested in the reaction at one electrode, an ideal reference electrode will have a relatively fixed potential regardless of the environment to which it is exposed. Thus changes of potential are attributed solely to the electrode under study and not the reference electrode.

The Ag/AgCl reference electrode consists of silver in contact with the insoluble salt AgCl. This couple is then in contact with a solution containing chloride ions. The reaction is then

$$Ag + Cl^- = AgCl(s) + e \tag{2}$$

Similarly the calomel half-cell consists of mercury in solution with Hg_2Cl_2 and chloride ions. The reaction is

$$Hg + Cl^- = \tfrac{1}{2}Hg_2Cl_2 + e \tag{3}$$

For the saturated calomel, Hg_2Cl_2 is ground up with solid KCl, producing a saturated solution of KCl.

However, it is most uncommon for electrochemical reactions to occur at standard conditions. More frequently, the activity of various species departs significantly from unity and the temperature varies from ambient. Utilizing thermodynamic fundamentals and the consideration of equilibrium conditions, we can derive an equation expressing the EMF of a cell in terms of the activities of products (a_p) and reactants (a_r) not in their standard states and at different temperatures:

$$E = E^\circ - \frac{RT}{nF} \ln \frac{a_p}{a_r} \tag{4}$$

where E° is the standard electrode potential of the cell, R is the universal gas constant, n is the number of electrons transferred, and T is the absolute temperature. This equation is referred to as the *Nernst equation.*

Thus for the oxidation of Cu,

$$Cu = Cu^{2+} + 2e \tag{5}$$

$$E_{Cu} = E^\circ_{Cu} - \frac{RT}{2F} \ln \frac{Cu^{2+}}{Cu} \tag{6}$$

where E°_{Cu} is the standard potential for the copper oxidation reaction. From equation (6) it is obvious that for any concentration of Cu^{2+} and any temperature the potential E_{Cu} can be determined.

From thermodynamics and half-cell potentials, it can be stated that corrosion will not occur unless the spontaneous direction of the reaction indicates metal oxidation. However, care must be taken since the reverse statement is not necessarily true. Just because the spontaneous direction is toward metal oxidation, this does not always mean that corrosion will proceed at a observable rate. Thermodynamics is concerned with direction, not rate.

Electrode potentials described by the Nernst equation are measured for the case of zero current flow between the metallic electrode and the standard hydrogen electrode. The potential measured is a reversible potential that reflects the maximum driving force for the reaction. Regardless of the direction in which the reaction is written, the magnitude of the potential remains constant; only the sign changes. If the reaction is written

$$M = M^{2+} + 2e \tag{7}$$

then oxidation of the pure metal is occurring with the release of electrons. In every case, the number of electrons produced equals the valence state of the metal ion created. Likewise, the reaction may be reversed to create a reduction reaction:

$$M^{2+} + 2e = M \tag{8}$$

Oxidation is the reaction associated with the anode (corrosion), while reduction is the cathodic reaction. Because of the electron transfer involved in these reactions, the two reactions depend on each other. However, the reaction rate involves kinetics, which will be discussed later.

One of the most widely used applications of thermodynamics to corrosion theory is in the form of *E_H-pH diagrams* or more commonly, *Pourbaix diagrams* named for their originator Marcel Pourbaix. These diagrams present the potential (E_H) of a metal electrode referenced to the standard hydrogen electrode based on Nernst equation calculations versus the pH of the aqueous solution in which the corrosion is taking place. To construct one of these diagrams, the following reactions must be considered.

- ☐ Electrochemical reactions involving only charge transfer (electrons)
- ☐ Electrochemical reactions involving electrons and H ions
- ☐ Chemical reactions

Electrochemical reactions involving only the transfer of electrons will produce a straight horizontal line on the E_H-pH diagram that is independent of pH. Likewise, strictly chemical reactions will produce a vertical line independent of E_H. Reactions involving both electrochemical and chemical dependence will generate sloping lines.

Figure 5 is an example of an E_H-pH diagram for iron in water at 25°C [3]. Line A represents electrochemical equilibrium between Fe^{2+} and Fe^{3+} ions with no dependence on the solution pH. The Nernst equation for this condition is

$$E_H = 0.771 + 0.0591 \log \frac{a_{Fe3+}}{a_{Fe2+}} \tag{9}$$

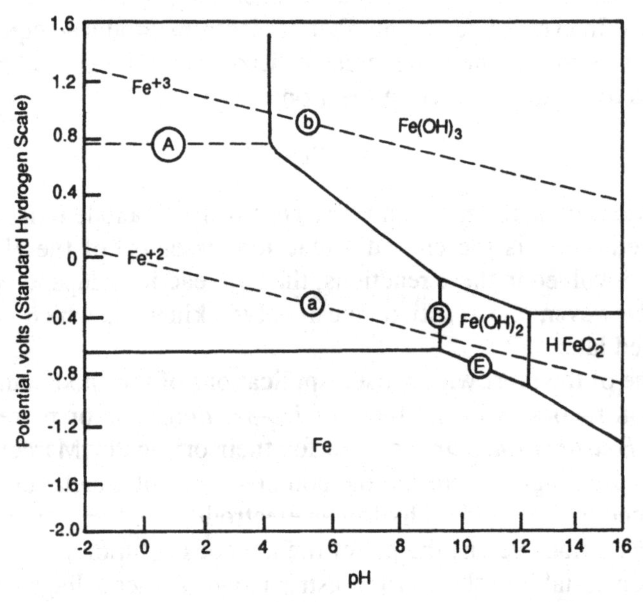

Figure 5. E_H-pH diagram for iron in water at 25°C.

Above this line is the region of thermodynamic stability for Fe^{3+} ions; here, iron will tend to corrode. Likewise below line A, Fe^{2+} ions are stable, and iron will still be susceptible to corrosion but produce ferrous rather than ferric ions. On the line itself ferrous and ferric ions are in equilibrium.

The activity (concentration) of each ionic species must either be known or assumed, to calculate the necessary conditions for an E_H-pH diagram. Concentrations of 10^{-6} M are often assumed for construction of these diagrams. However, quite often several concentration lines are presented on a diagram for more information.

Line B is the boundary that describes purely chemical equilibrium independent of potential between ferrous ions and ferrous iron hydroxide, $Fe(OH)_2$. This behavior is derived from the chemical equation:

$$\log K = \frac{\Sigma \nu_R \mu_R^\circ - \Sigma \nu_P \mu_P^\circ}{2.3RT} \tag{10}$$

where K is the equilibrium constant, μ° is the standard chemical potential of either the products (P) or reactants (R), and ν is the stoichiometric coefficients of the products (P) or reactants (R).

For these chemical reactions, there are no dependencies on potential, but they are highly dependent on pH. To the right of line B, $Fe(OH)_2$ is stable and theoretically provides a reaction product (corrosion film) that reduces corrosion further by interference of the electrochemical reactions at the metal surface. Therefore, regions that allow for thermodynamically stable compounds are designated as regions of passivity. The concept and reality of passivity are much more complicated than implied by the Pourbaix diagram. These aspects will be explored in greater detail in this book. To the left of line B, the ionic species Fe^{2+} is stable, and corrosion will continue unabated with no formation of a stable corrosion film.

Lines a and b indicate reactions that depend on electrochemical potential and pH. These lines represent the regions of stability for water. Above line b the evolution of oxygen occurs by the decomposition of water according to

$$H_2O \rightarrow \tfrac{1}{2}O_2 + 2H^+ + 2e \tag{11}$$

Below line a, evolution of hydrogen occurs:

$$2H^+ + 2e \rightarrow H_2 \quad (12)$$

Within the boundaries of these lines water is stable. The importance of water stability/instability is so necessary to a good understanding of corrosion behavior of metals that these lines are almost always present as dashed lines on all Pourbaix diagrams. Line a is included on the dashed line for the reduction of hydrogen, and line b for the evolution of oxygen.

Line E represents equilibrium between $Fe(OH)_2$ and solid iron (Fe). This equilibrium is both potential and pH dependent. Above the line, stable $Fe(OH)_2$ creates a region of passivity, whereas below the line iron metal is stable and does not corrode. This latter region is referred to as a region of *immunity*. Immunity for iron in water at 25°C is pH independent below about pH 9 and, therefore, is strictly an electrochemical reaction.

Figure 6 is another method of displaying Pourbaix diagrams to emphasize those regions of expected corrosion and those for which corrosion is either mitigated by a passive reaction film (passivity) or not allowed at all by thermodynamics [3].

Pourbaix diagrams have found extensive use in corrosion science, as well as in areas such as geology and geochemistry. These diagrams provide a useful tool for predicting corrosion of metals or alloys in a variety of environments. Furthermore, they may be used for predicting changes that can be applied to a system to reduce

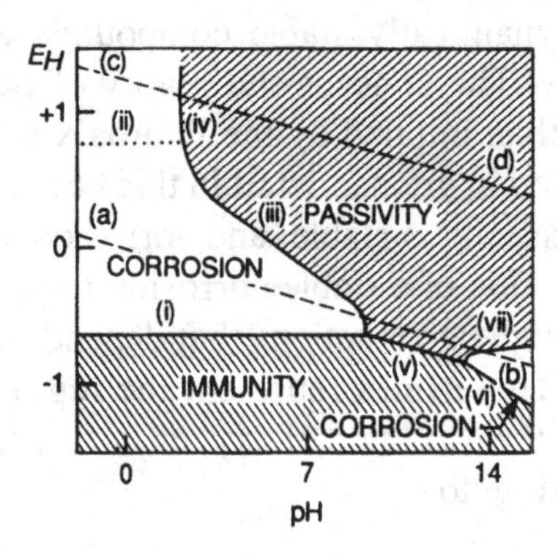

Figure 6. Simplified E_H-pH diagram of Fig. 5.

corrosion, as in the case of cathodic protection, and for estimating the composition of corrosion products. It must be emphasized that these diagrams are thermodynamic in nature and cannot be extended to kinetics. The composition of compounds in each diagram represents the equilibrium composition, whereas in many systems where corrosion occurs the nature of the corrosion product is transitional and may not appear on any E_H-pH diagram. Also, in many systems duplex and triplex oxides form that are not predicted in the diagram. Therefore, these diagrams are suitable only as guidelines or for a system that is known to be at thermodynamic equilibrium.

Few E_H-pH diagrams exist for alloys, which are more complicated yet more commonly used than pure metals, and most corrosive environments do not contain just water. The inclusion of other chemical species and alloys, while more realistic from an application viewpoint, does make the generation of such diagrams much more complex. However, with the extensive use of computers, Pourbaix diagrams for many systems of importance can be generated. Likewise, temperature has a profound effect on corrosion. Figure 7 illustrates the benefits of a three-dimensional plot of E_H, pH, and T [4]. It can be seen that several corrosion films will be in equilibrium, de-

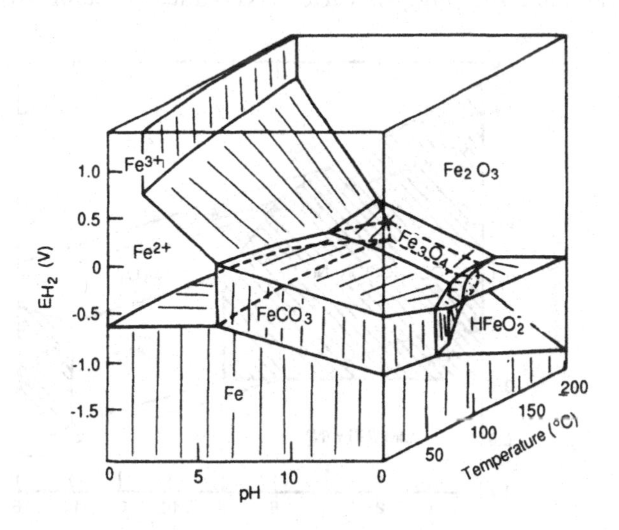

Figure 7. E_H-pH-temperature diagram for the Fe–CO_2–H_2O system. Partial pressure $CO_2 = 3.0$ MPa and activities of dissolved species at 10^{-6} M.

pending on temperature. For example, Fe_3O_4 and $FeCO_3$ may coexist as well as Fe_2O_3 and Fe_3O_4.

One further drawback of these diagrams is that the pH used in calculations is theoretically the pH measured immediately adjacent to the electrode surface (i.e., the electrode interface). However, it is currently impossible to measure the actual pH adjacent to a metal surface; thus, bulk pH is used. In many corrosive systems the bulk pH is entirely different from the pH immediately adjacent to the charged surface. In later chapters it will be shown that the pH at the base of pits and crevices or at the tip of an advancing stress corrosion crack is often many pH units lower compared with the bulk. In these cases, the bulk pH would not correctly predict the position on a Pourbaix diagram. Additionally, the pH adjacent to a cathodic area is often much more caustic than the bulk solution pH.

In regard to the focus of this book, Pourbaix diagrams offer the first glimpse of the role of surface films in reducing or eliminating corrosion, thereby emphasizing the importance of surface films in corrosion science.

When other metal systems are examined in different electrolytes, broad regions of stable corrosion films are often evident. For instance, Ni in water at 25°C (Fig. 8). Various Ni oxides are stable over a range

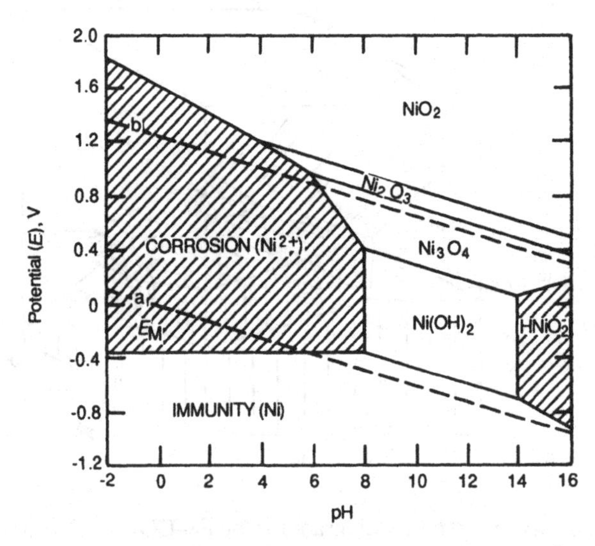

Figure 8. E_H-pH diagram for nickel in water at 25°C.

of potentials. However, NiO_2 has the widest pH stability. Yet at low pH and high pH Ni is susceptible to corrosion. Additionally, it is evident that several of the oxides can coexist at specific E_H-pH values so that a duplex oxide will be stable. For example, at pH 11 and 0.95 V, NiO_2 and Ni_2O_3 would be present in equilibrium with each other. This is commonly observed in practice for many metallic systems that form oxides and sulfides in much more complex corrosive environments.

More noble metals such as Pt display much larger regions of immunity and passivity compared to more active metals such as iron (Fig. 9). Figure 10 shows that the domain of immunity almost covers the region of stability for water. Thus platinum is stable in aqueous solutions for essentially all pH's at 25°C. This diagram, compared to those for Fe and Ni, illustrates the advantage of Pourbaix diagrams in predicting from thermodynamic data which environments may produce corrosion and which metals should be resistant.

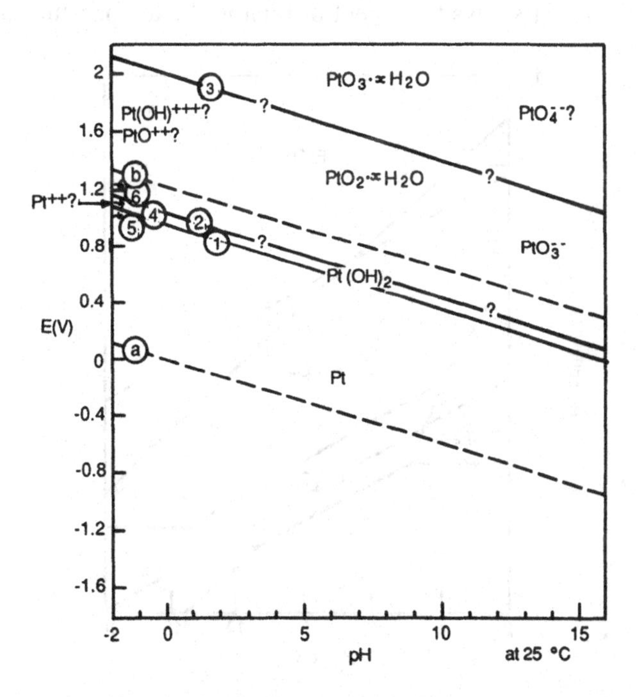

Figure 9. E_H-pH diagram for platinum in water at 25°C.

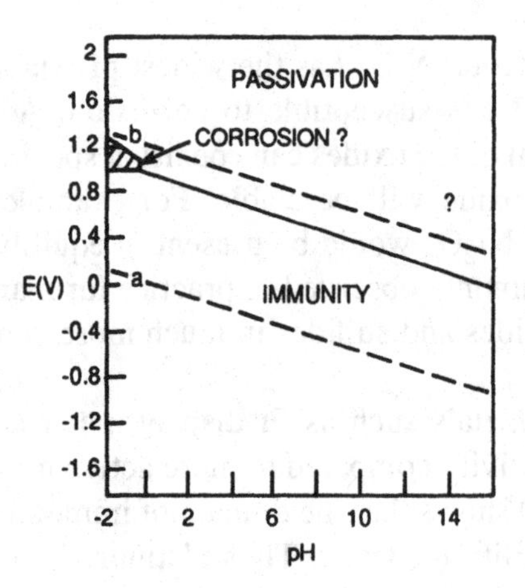

Figure 10. Simplified diagram of Fig. 9.

As previously mentioned, most of these diagrams have been generated for pure metals at ambient temperature (25°C) and pressure (1 atm). Figure 11 shows the effect of raising the temperature for the

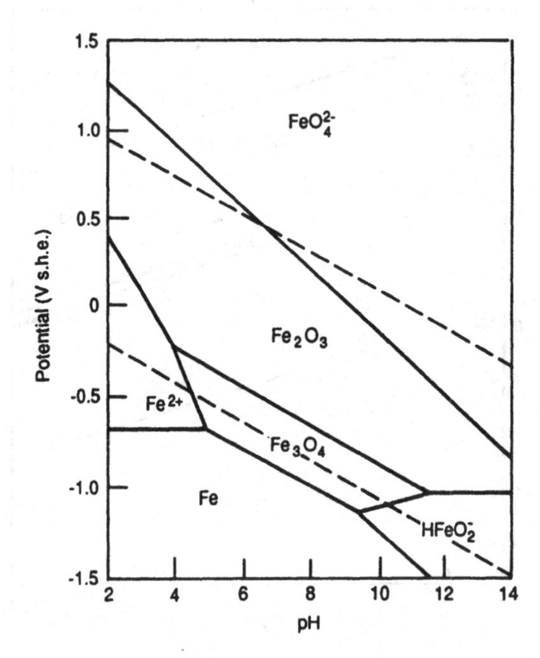

Figure 11. E_H-pH diagram for Fe in water at 250°C.

Fe-H_2O system to 250°C [5]. Compared to Fig. 5, several changes are apparent. At 250°C, ferric ions are no longer stable at any combination of pH and potential. The immunity region has become smaller, but regions of passivity have grown due to the greater stability of certain oxides (i.e., Fe_2O_3) at higher temperatures.

Changes in the environment can likewise significantly alter the passivity domains and the composition of those films that are stable under other conditions. Figure 12 is the Pourbaix diagram for the Fe/H_2O system with 0.1 *M* at-g sulfur [6]. Compared to Fig. 5, certain sulfides are more stable than oxides, and *vice versa,* depending on potential and pH. The regions of stability for ferrous and ferric ions (corrosion) have been compressed to pH values of 2 or less, compared

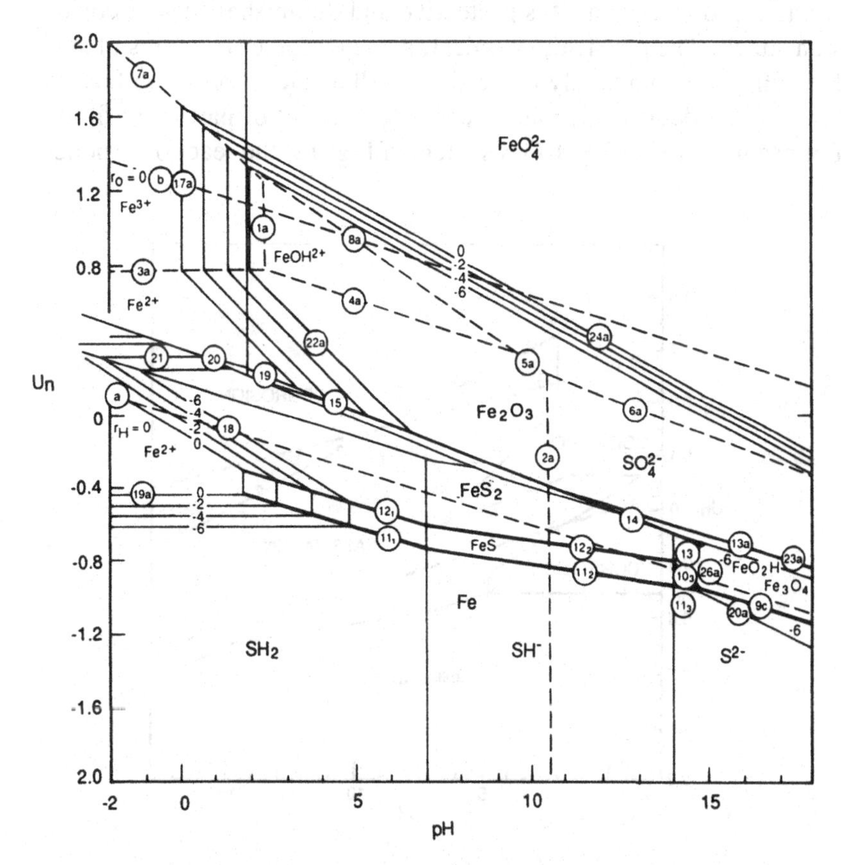

Figure 12. E_H-pH diagram for iron in water at 25°C with 0.1 *M* at-g S.

with almost pH 9 in Fig. 5. The region of immunity has remained essentially constant. In addition to the compounds formed, this figure presents the stable anion species in each pH/E_H domain (i.e., sulfate, sulfides, etc.).

Figure 13 is presented for ease of comparison to Fig. 6. Much of the passivity at more positive potentials in Fig. 6 has disappeared in Fig. 13 due to the introduction of a small concentration of sulfur. This example illustrates the profound effect small concentrations of impurities in an environment may have on both the composition of the corrosion film and on the susceptibility to corrosion.

Yet, Pourbaix diagrams oversimplify the case for oxide or sulfide films formed on metals. It is implicit in these diagrams that if a reaction product forms it is protective and diminishes further corrosion attack. The physical, chemical, and electronic structures of surface films are completely ignored, as well as the effect these factors have on the degree of protectiveness (passivity) of particular films. For example, in the Fe/H_2O/S system of Fig. 12, the reaction product

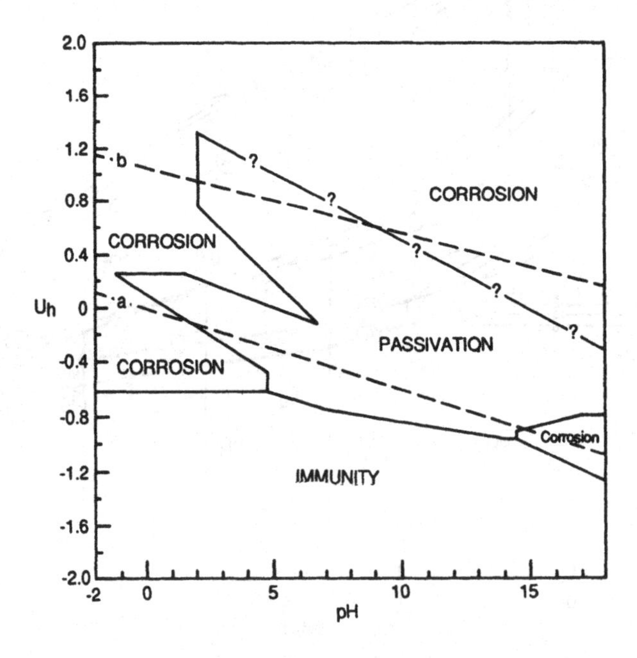

Figure 13. Simplified diagram of Fig. 12.

FeS is stable over a large range of pH but a small potential range. However, numerous ferrous sulfides with the general chemical formula FeS vary considerably in crystal structure and stoichiometry. The iron-rich phase $Fe_{1+x}S$ (mackinawite) is the most commonly formed corrosion product in H_2S/Fe systems at ambient temperature. Two other less common phases that can form are, the sulfur excess phase $Fe_{1-x}S$ (pyrrhotite) with a composition from $x=0$ to $x=0.14$, and the stoichiometric phase FeS (troilite). These nonstoichiometric films can have a profound impact on the protectiveness of the corrosion film (reaction product), contrary to the E_H-pH diagram prediction of passivity.

High-Temperature Aspects

In the absence of liquid water at higher temperatures, reaction of the environment, especially gases, will also depend on the thermodynamics of the system. Analogous to aqueous corrosion, the thermodynamics of high-temperature corrosion are described by the equation

$$\Delta G = \Delta G^\circ + RT \ln \frac{a_p}{a_r} \tag{13}$$

For example, the oxidation of a metal would be

$$xM + (y/2)O_2 = M_xO_y \tag{14}$$

where M is the metal, M_xO_y is the metal oxide formed, and x and y are the moles of metal and oxygen, respectively. The Gibbs free-energy change or driving force for the oxidation reaction would then be

$$\Delta G = \Delta G^\circ + RT \ln \frac{a_{M_xO_y}}{(a_M)^x (a_{O_2})^{y/2}} \tag{15}$$

The activities of solids (metals and oxides) by convention are taken as equal to 1. Thus, at equilibrium ($\Delta G = O$) the equation reduces to

$$\Delta G^\circ = (y/2)RT \ln a_{O_2} \tag{16}$$

For many cases of high-temperature oxidation where the temperatures are relatively high and the partial pressures of gases are moderate, the activity of oxygen can be approximated by its pressure P_{O2}. The equation then becomes

$$\Delta G^\circ = (y/2)RT \ln P_{O_2} \tag{17}$$

Thus, like the equations developed for electrochemical potential dependence on the hydrogen ion activity, so also high-temperature oxidation is found to depend on the concentration (pressure) of the oxidizing species.

A common means of presenting data on the stability of oxides in equilibrium with their metals is shown in Fig. 14, called an *Ellingham* or *Richardson diagram* [7]. The inflections in certain straight lines in the diagram represent a phase change for the liquid oxide going to a gaseous phase.

The standard free energy of formation for any oxide (ΔG^O) can be determined from this diagram: the more negative the free energy, the more thermodynamically stable the oxide. For example, NiO has a free energy of formation of about -300 kJ at 800°C for 2 moles NiO per mole of oxygen gas. Conversely, the equilibrium partial pressure of oxygen can be determined by utilizing the P_{O2} scale to the bottom and right of the figure. If we place a straight line from the *O* on the left-hand scale of the diagram through the 800°C/NiO intersection and continuing over to the right-hand scale, the line will intersect the P_{O2} scale at the equilibrium oxygen pressure. The value obtained is approximately 10^{-14} atm. Thus oxygen pressures greater than this value will tend to oxidize Ni, and those less than 10^{-14} atm will tend to decompose (reduce) NiO to Ni plus oxygen if the oxide is already present. For the pure metal, oxidation will not occur below this pressure.

Two other environments, H_2O and CO/CO_2, will also produce oxidation of metals. Some diagrams contain axes for both, but Fig. 14 only presents the CO/CO_2 case. The same method is employed, but the index point, C, for CO/CO_2 is used on the left scale as required.

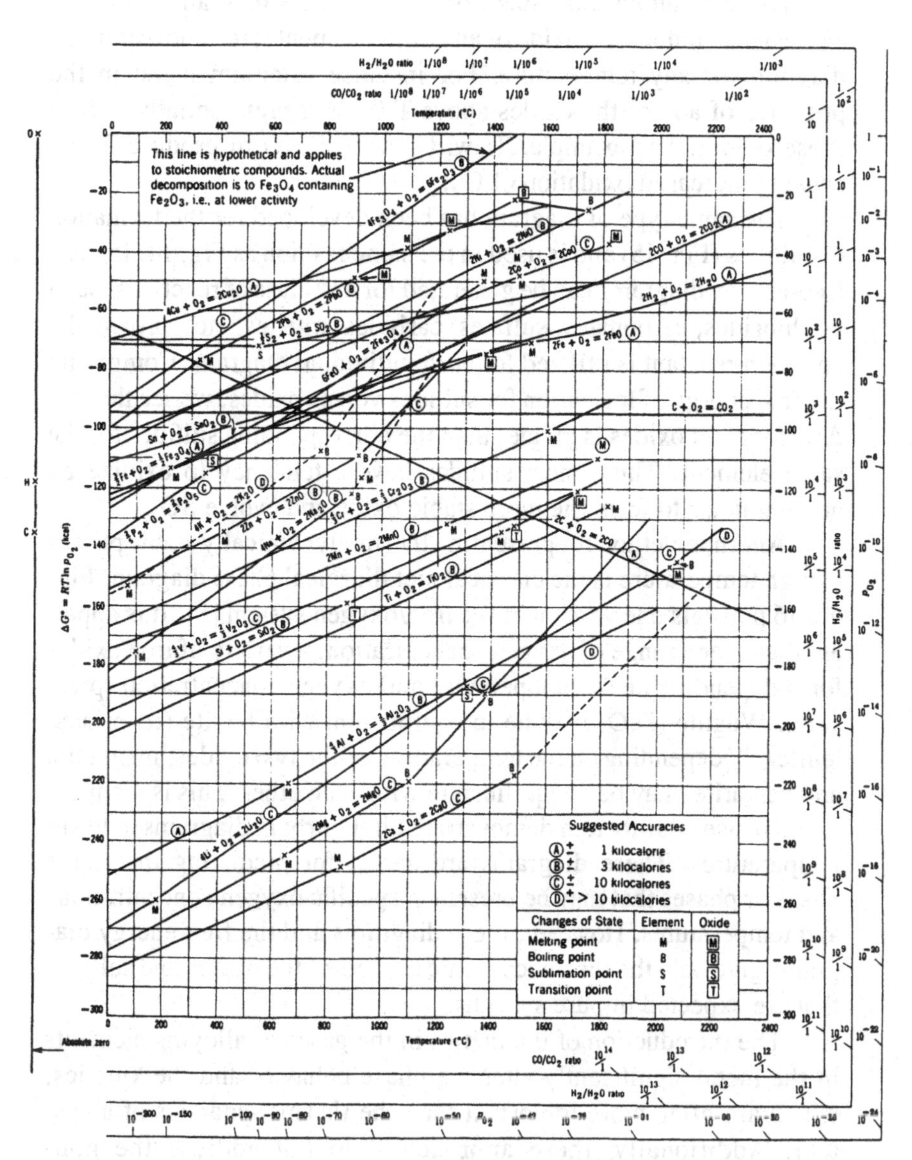

Figure 14. Standard Gibbs free energy of formation of oxides as a function of temperature.

Another important aspect of this versatile diagram is that an element can reduce the oxide of any other element lying above it in the diagram for any temperature. For instance, titanium metal in the presence of any of the oxides above TiO_2 will preferentially reduce those oxides. For example, Ti will reduce Cu_2O to produce TiO_2, causing increased oxidation of Cu.

The same type of diagram has been developed for the formation of sulfides (Fig. 15) and is used in the same fashion as Fig. 14. In fact, free-energy diagrams can be generated for any class of reactions, such as chlorides, carbonates, sulfates, carbides, nitrites, etc., provided a common reactant is utilized for each individual diagram. Comparing the free energy of formation for sulfides versus oxides, we see that the $\Delta G°$ for most oxides is greater than the corresponding sulfides for the same element. Thus, most sulfides have a tendency to oxidize by heating in air to form the more stable oxide and evolve SO_2.

Another method for presenting thermodynamically stable phases at high temperature is the classical metallurgical phase diagram. Figure 16 presents one version of the iron/oxygen system [8]. It is apparent that over a range of oxygen concentration, a variety of iron oxides form depending on the temperature and oxygen concentration (pressure). Wustite (FeO) may be in equilibrium with ferrite (α) or austenite (γ), depending on the temperature, or the two oxides, magnetite and hematite, may be in equilibrium with each other. This is identical to the phase equilibrium demonstrated in Pourbaix diagrams at lower temperatures. These diagrams are useful for predicting the stable phase or phases that may be present at specific oxygen concentrations and temperatures. However, these diagrams and the free-energy diagrams give only the thermodynamic aspects of the oxides and sulfides that are expected in pure systems.

The introduction of impurities in the gases or alloying elements in the metal significantly alter the phase behavior and the kinetics, which are often more important than the thermodynamics of a system. Additionally, these approaches do not address the nonstoichiometry of the oxides or sulfides that form. Other properties of the scale are not presented that contribute significantly to the protectiveness of the oxide or sulfide and its growth. Many of these factors will be discussed in later chapters.

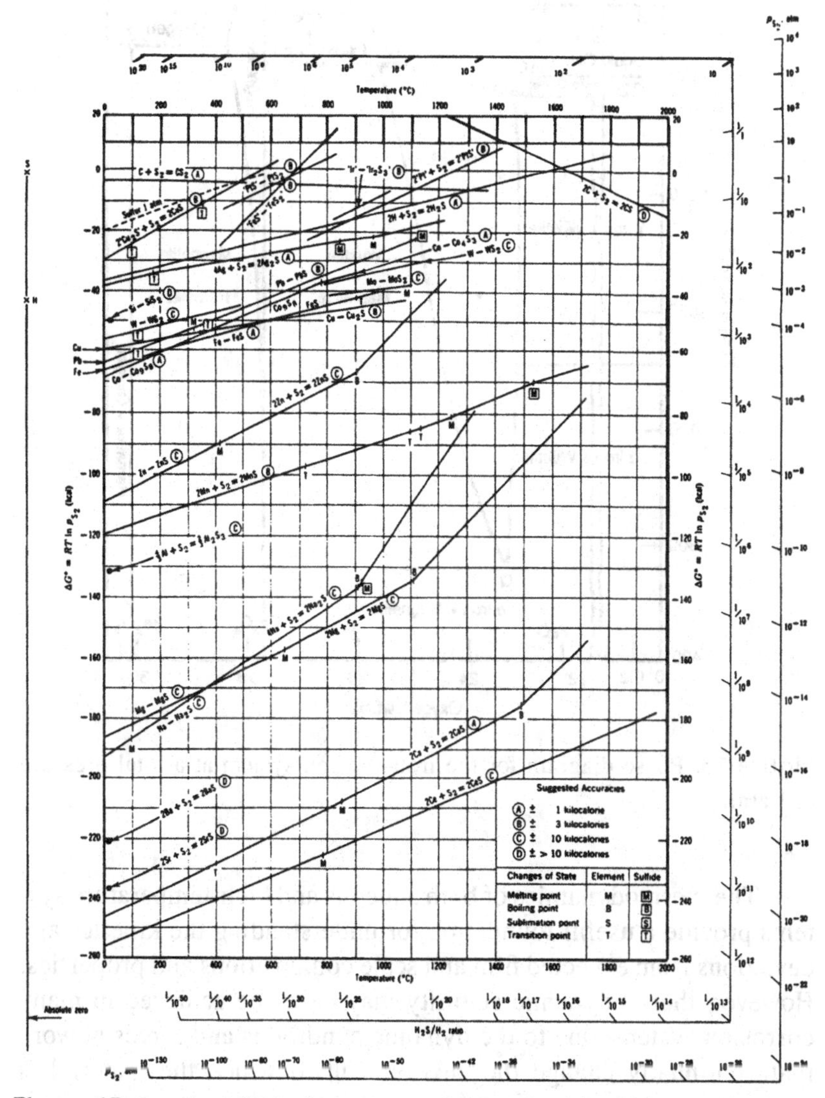

Figure 15. Standard Gibbs free energy of formation of sulfides as a function of temperature.

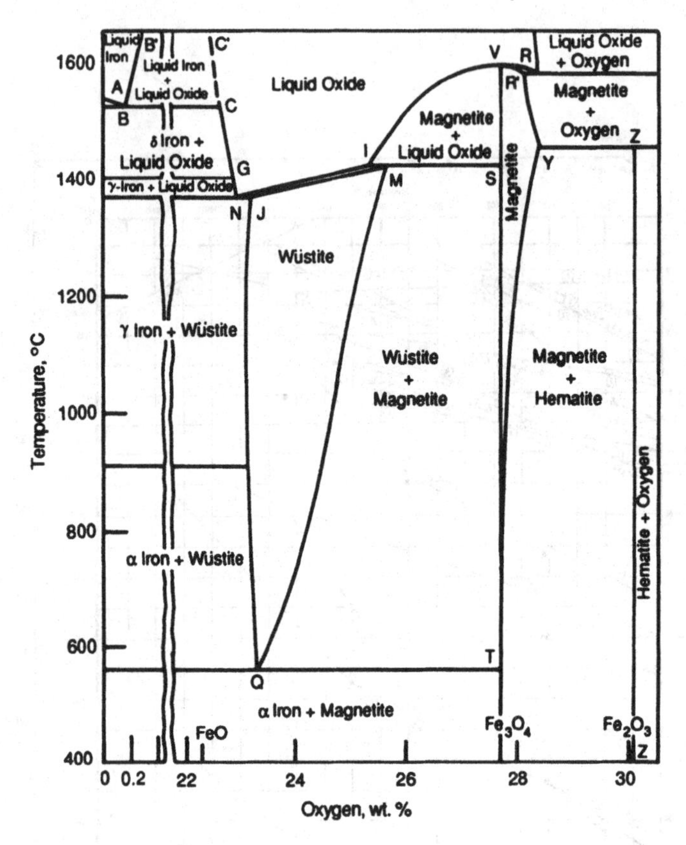

Figure 16. Phase diagram for the iron–oxygen system at a total pressure of 1 atm.

The thermodynamics of both aqueous and high-temperature systems provide a useful foundation for understanding the kinetics and deviations from expected film and scale compositions and properties. However, thermodynamic stability may never be achieved in many corrosion systems due to the dynamic conditions and forces at work that continually change the environment to which the material is exposed.

Care must always be taken in applying thermodynamic data especially from a design standpoint and should be complemented with kinetic data to fully understand the complete corrosive environment and the behavior of a metal or alloy exposed to it.

References

1. J.O'M. Bockris and A.N. Reddy, *Modern Electrochemistry,* Plenum, New York, 1977.
2. Corrosion, vol. 13, *ASM Metals Handbook,* ASM International, 1987.
3. M. Pourbaix, *Atlas of Electrochemical Equilibria,* NACE, 1974.
4. A. Ikeda, M. Ueda, and S. Mukai, Corrosion/83, Paper No. 45, NACE, Anaheim, CA, 1983.
5. C. M. Chen, K. Aval, and G. J. Theus, EPRI Report NP-3137, Palo Alto, CA, 1983.
6. J. Bouet and J.P. Brenet, *Corrosion Science* **1963,** *3,* 51.
7. L.S. Darken and R.W. Gurry, *Physical Chemistry of Metals,* McGraw-Hill, New York, 1953.
8. *The Making, Shaping and Treating of Steel,* U.S. Steel Corp., 1979.

2

Kinetics of Corrosion

Introduction

As described in Chapter 1, there is a thermodynamic driving force for corrosion to proceed. The oxidation and reduction reactions for a corrosion cell were presented as independent processes. However, they depend on each other. For instance,

$$Fe \rightarrow Fe^{2+} + 2e \qquad \text{(anodic)} \tag{1}$$

$$2H^{+} + 2e \rightarrow H_2 \qquad \text{(cathodic)} \tag{2}$$

The two electrons in both equations are the current that generates the electrochemical reaction. Since, by the law of conservation of charge, no electrical charge can accumulate at the anode or cathode, the total current flowing out of the anodic reaction must equal the total current flowing into the cathodic reaction. To maintain the mass balance portion (chemical) of these related reactions, the amount of cathodic constituents consumed must be equal to the amount of corrosion product formed. Since these two reactions depend on each other, an increase or decrease in the rate of one will likewise increase or decrease the other. Thus, the introduction of a cathodic depolarizer (enhances the cathodic reaction) will in turn increase the anodic reaction (corrosion) rate.

The combination of two or more electrodes (i.e., iron, hydrogen) allowing current to flow between them results in a deviation from the equilibrium potential for each electrode. This deviation from equilibrium is termed *polarization,* and the magnitude of deviation is referred to as *overvoltage,* expressed as η. Polarization is simply the resulting electrode potential produced by a net current flow. Overvoltage is the measured deviation in potential, assuming the equilibrium potential as zero. Most often, overvoltage is used in conjunction with the hydrogen or oxygen electrode reaction.

At equilibrium the oxidation rate is equivalent to the reduction rate; therefore, the current exchange between the anode and cathode is constant but not necessarily zero. This current, or more conveniently current density, is defined as the exchange current density i_0. The iron electrode at equilibrium is shown in Fig. 1. Here the reduction of iron is equal to the oxidation of iron.

The exchange current density of an equilibrium electrode is a strong function of the electrode composition, surface roughness of the electrode, the presence of trace impurities, temperature, and the ratio of oxidized and reduced species present. Therefore, the exchange current density for an oxide-covered electrode will be significantly different than that for a true bare electrode.

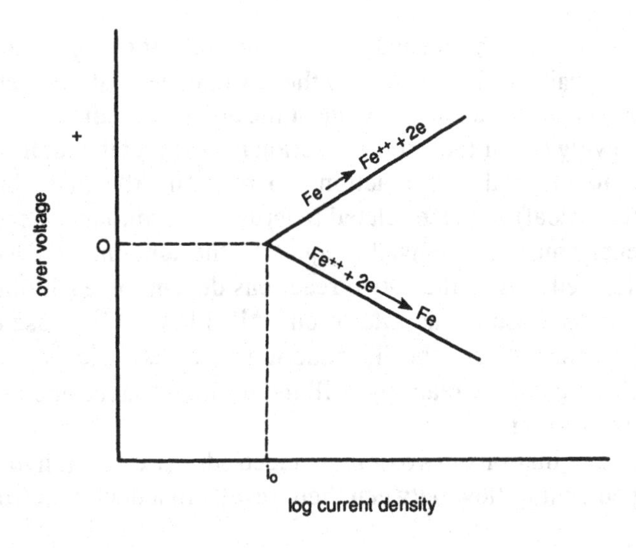

Figure 1. Exchange current density for the oxidation-reduction of iron.

Polarization

Polarization is typically divided into *activation* polarization and *concentration* polarization. Activation polarization requires an activated step to occur for the reaction to proceed. A good example is the reduction of hydrogen ions at a cathode. The slowest step is the rate-determining step, which is either the electron transfer at the electrode solution interface to produce hydrogen atoms or the formation of hydrogen molecules from the hydrogen atoms residing at the electrode surface.

There are two components of the activated step: chemical and electric. The movement of hydrogen ions from outside the outer Helmholtz plane to the electrode surface requires surmounting a potential energy barrier (Fig. 2), which is strictly a chemical phenomenon [1]. In the presence of an electric field, an additional component of electrical work must be included with the chemical contribution. Likewise in the case of the anodic process, the release of, for example,

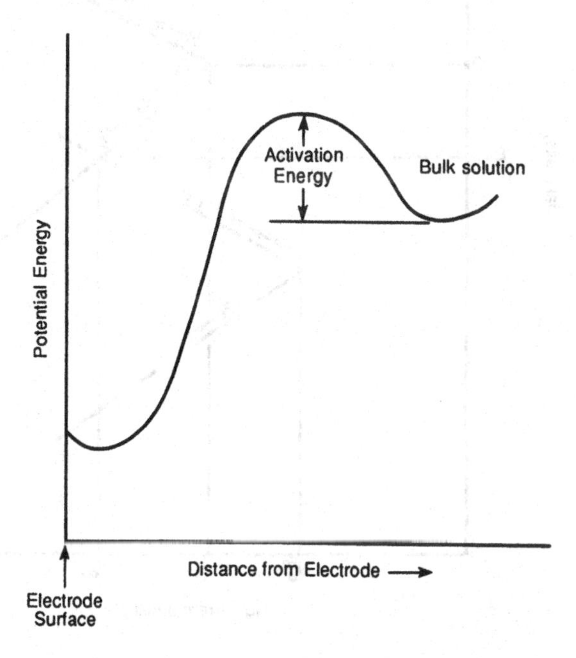

Figure 2. The activated step for transport of an ion in the bulk solution to the electrode surface.

ferrous ions from an iron electrode will require surmounting an activation barrier before Fe^{2+} can be released from the interface and be transported into the solution. These activated processes, and therefore the activation polarization, are often ideally described by an Evans diagram, previously shown in Fig. 1. The combined cathodic and anodic reactions of hydrogen and iron just mentioned can be easily illustrated using this type of diagram. Figure 3 shows the combination of these two processes and the resulting corrosion parameters, E_{corr} and i_{corr}. E_{corr} is the mixed potential or corrosion potential, and i_{corr} is the corrosion current density. The sloping lines

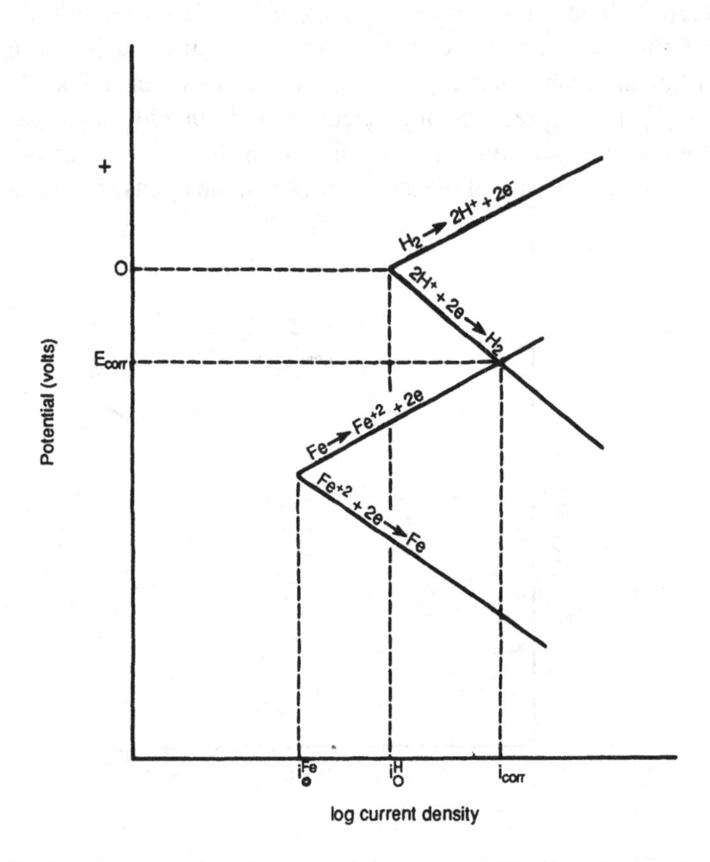

Figure 3. The development of a corrosion potential (E_{corr}) and corrosion current (i_{corr}) from the combination of two equilibrium electrodes.

thus represent the activated processes for both the forward and reverse reactions for the iron electrode and the hydrogen electrode.

This departure from equilibrium at the interface can be described mathematically by the Butler-Volmer equation

$$i = i_0\left[\exp\left[\frac{(1-\alpha)}{RT}\eta F\right] - \exp\left[\frac{-\alpha\eta F}{RT}\right]\right] \tag{3}$$

where η is the overpotential, $E - E^{\circ}$, R is the gas constant, T is the absolute temperature, i_0 is the equilibrium exchange current density, α is a symmetry coefficient, and F is Faraday's constant. The symmetry coefficient is a factor greater than zero but less than one and is equivalent to the distance across the double layer to the top of the activation peak, illustrated in Fig. 2, divided by the entire double-layer width. For simplicity, the activation energy curve is assumed to be symmetric; thus $\alpha = 0.5$.

When a positive ion such as hydrogen is transferred to the electrode (reduction), it must do electrostatic work against the double-layer field, or, in other words, the field does work on the ion. This work on the positive ion is equivalent to reducing the magnitude of the activation energy barrier and is described by the factor $\alpha\eta F$ in equation (3). Conversely, for the transfer of an ion from the metal to the solution (oxidation), the activation barrier increases by the amount $(1-\alpha)\eta F$.

The current density across the metal/solution interface is strongly dependent on overvoltage η, and small changes in η produce significant changes in i. At equilibrium $\eta = 0$ and $E = E^{\circ}$. No measurable current flows, but since this is a dynamic equilibrium the rate of oxidation equals the rate of reduction:

$$i_a = -i_c = i_0 \tag{4}$$

where $E > E^{\circ}$ oxidation proceeds opposing the cathodic reaction, and *vice versa* when $E < E^{\circ}$. As the overpotential becomes sufficiently large, one reaction overwhelms the other. For $E - E^{\circ} >> 0$,

$$i_a = i_0 \exp\left[\frac{(1-\alpha)\,\eta F}{RT}\right] \tag{5}$$

and for $E - E^{\circ} << 0$,

$$i_c = i_0 \exp\left[\frac{\propto \eta F}{RT}\right] \tag{6}$$

At the high values of η where these equations are valid, activation polarization is no longer the controlling mechanism of polarization. Rather, polarization becomes concentration dependent and is termed *concentration* polarization. This phenomenon is created by the concentration gradient produced across the double layer by reaction of the ions at the electrode surface. As the electrode reaction increases under activation control, the double layer becomes depleted in ions available for reaction, thereby producing a gradient across the double layer. As the concentration gradient increases, the ability to supply ions from the bulk solution to the electrode becomes rate-limiting and concentration polarization will be the rate-controlling step.

To better illustrate this behavior, the hydrogen reduction reaction described before is utilized. Under low reduction rates, the distribution of hydrogen ions in the electrolyte will be relatively uniform. As the overpotential increases, the reduction rate of hydrogen ions to hydrogen atoms and thence to hydrogen gas will increase rapidly, and the concentration of hydrogen ions adjacent to the electrode surface will be diminished. Coincidentally, a concentration gradient will occur across the double layer extending into the bulk solution. Eventually, the rate at which hydrogen ions can be supplied to the electrode surface by diffusion will become limiting, and further increases in potential will not produce additional increases in current because of the lack of hydrogen ions at the interface for charge transfer. This limited current is referred to as the *limiting diffusion current density*, i_L, given by

$$i_L = \frac{DnFC_0}{x} \tag{7}$$

where D is the diffusion coefficient of the reacting ions, n is the number of electrons transferred, F is the Faraday constant, C_0 is the concentration of reacting ions in the bulk solution, and x is the thickness of the diffusion layer. This limited diffusion current density represents the maximum rate of reduction possible in a given system. Limiting current density is only important in reduction processes and is usually negligible in metal dissolution reactions, so it can be ignored in corrosion reactions.

Both activation and concentration polarization typically occur at the same electrode. Combining the two (Fig. 4), we see that activation polarization is predominant at low reduction rates, while concentration polarization is controlling at higher reduction rates.

A physical model of the diffusion layer is presented in Fig. 5 [2]. The solid line is derived from Fick's first law of analysis for steady-state diffusion given the initial bulk concentration of reacting species, C_0^b and the initial electrode concentration of the same species, C_0^s. The dashed line represents a simplified approach in accordance with equation (7). The thickness of the diffusion layer remains relatively constant in the presence of natural convection. However, if diffusion is

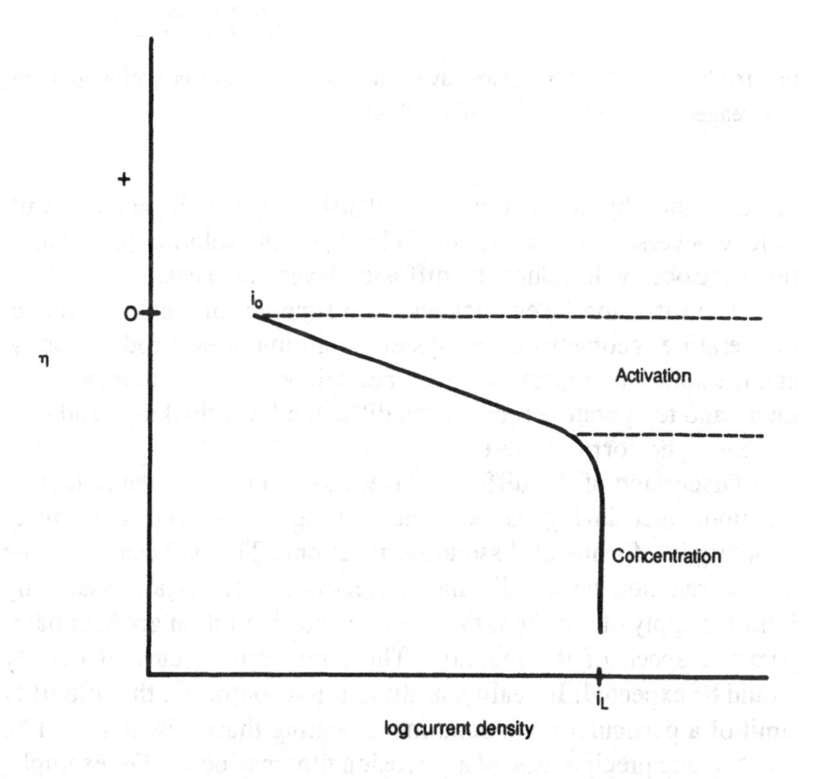

Figure 4. Potential/current regions for which activation polarization and concentration polarization control the electrode process.

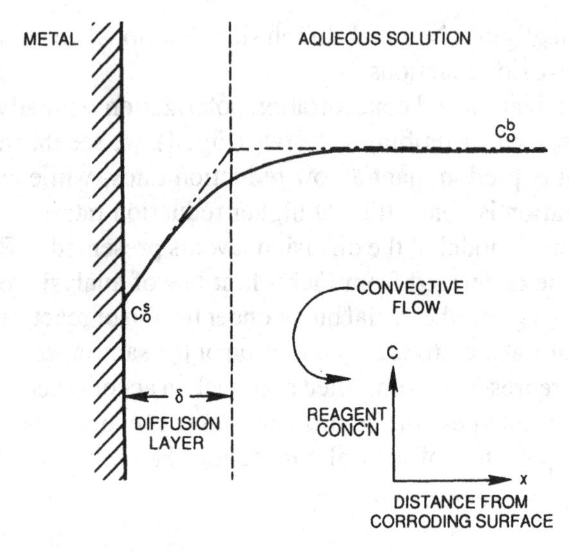

Figure 5. The Nernst diffusion layer illustrating the concentration gradient of a reagent O in the vicinity of a cathode.

not disturbed by convection, the diffusion layer will increase with time. Conversely, increasing the velocity of the solution (or rotating the electrode) will reduce the diffusion layer thickness.

The diffusion layer thickness is a function of electrode shape, temperature, geometry of the system, solution or electrode velocity, and the solution properties (i.e., kinematic viscosity). Increasing velocity and temperature reduces the diffusion layer thickness and thus increases the corrosion rate.

Discussion of the diffusion layer has centered on the reduction reaction, since during oxidation the limiting diffusion current density is negligible for metal dissolution reactions. This is because in the anodic reaction on a solid metal electrode there is an essentially infinite supply of ions and, therefore, no concentration gradient hampers the speed of the reaction. Thus, no limiting current density would be expected. In reality, as dissolution continues, the solubility limit of a particular compound incorporating these new ions will be reached and precipitation of a corrosion film may occur. For example, during the oxidation of iron, iron hydroxide often precipitates on the electrode. This precipitation of a corrosion film will effectively produce a limiting current density for the anodic reaction. Transport of ionic species through the film may then become the limiting step.

If this limiting condition is ignored, the relationship between overpotential and current density for corrosion (i_{corr}) is given by

$$\eta_{oxi} = \beta \log \frac{i_{corr}}{i_0} \tag{8}$$

and that for reduction is

$$\eta_{red} = -\beta \log \frac{i_{corr}}{i_0} + 2.3\frac{RT}{nF} \log \frac{(1 - i_{corr})}{i_L} \tag{9}$$

where β is the Tafel slope. The Tafel slope represents the slope of the log i versus η curve of a multistep reaction at the metal–solution interface.

Equation (8), though frequently applied, is strictly valid only if no oxide films are present, if the IR drop in the solution is negligible (i.e., high conductivity), and the corrosion potential is sufficiently large for equation (5) to be applicable. However, very few cases actually meet all of these requirements, but equation (8) is often used as an approximation for the corrosion current (corrosion rate) because of its simplicity and ease of application.

The IR drop is another important consideration in the polarization behavior of electrodes. To this point, the resistance of the cell circuit has been assumed negligible. However, an ohmic potential drop exists through the double-layer electrolyte and any oxide films that are present. This contribution to polarization is equal to the product of the current density i and the resistance of the circuit. The major contributions to the circuit resistance are those described above and are equal to l/k, where l is the length of the current path and k is the specific conductivity. The inverse of the conductivity is resistivity, ρ, and the average resistivity is called the *faradaic resistance*. From Ohm's law, the total contribution to potential from these factors is

$$E = i\rho l \tag{10}$$

This equation assumes a linear potential difference exists with distance l to the reference electrode, which is not entirely accurate. Since many solutions have low resistivity and the resistances of corrosion films are often ignored, the IR contribution is frequently omitted. However, these contributions can be quite large for environments of low conductivity (i.e., pure water and nonaqueous electrolytes) or for corrosion films that behave as effective insulators.

If polarization resistance is measured at the corrosion potential E_{corr}, two contributions to resistance will be measured [3]: the ohmic resistance (IR drop) R_{Ω}, and the true polarization resistance R_p, so that the total resistance R_T is

$$R_T = R_p + R_{\Omega} \tag{11}$$

An electrical analog of this system is illustrated in Fig. 6. It has been found that the double layer and existing films will produce an equivalent electrical network that has both resistive properties R_{Ω} and capacitive properties C. The capacitance is given by

$$C = \frac{1}{2\pi R_p f} \tag{12}$$

where f is the frequency for the maximum imaginary impedance. The capacitance generated from this relationship is a measure of the double-layer capacitance and the thickness of the corrosion film on the electrode. The thickness of such a film can then be calculated by the relation

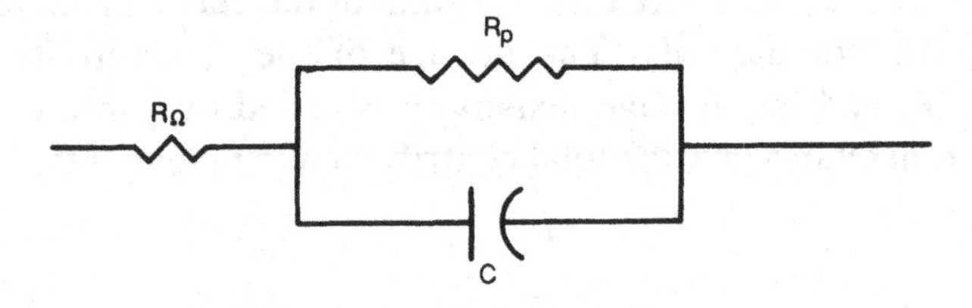

Figure 6. Electrical analog of the electrode/solution interfacial region.

$$t = \frac{A\epsilon}{C} \tag{13}$$

where t is the oxide film thickness, A is the true surface area of the electrode, C is the capacitance of the oxide, and ϵ is the electric permittivity of the film. Some difficulties of this relationship are determining the true area of the electrode, since surface roughness, pitting, and crevices will increase the area beyond the measurable area. Additionally, the electric permittivity of films is largely unknown. While permittivity is related to the dielectric constant K by

$$\epsilon = K\epsilon_0 \tag{14}$$

where ϵ_0 is the permittivity of free space (8.854×10^{-10} farad/m), very little data exist for the dielectric constant of corrosion films. Examples of some known dielectric constants for films are alumina (Al_2O_3), 4.5–8.4; cupric oxide (CuO), 18.1, and wustite (FeO), 14.2 [4]. The higher the dielectric constant is, the more a film behaves as an insulator.

By using an alternating current (AC), one can distinguish the true polarization resistance contribution to R_T in equation (11) from the ohmic resistance. The method, referred to as *AC impedance,* or electrochemical impedance spectroscopy, produces a plot which, in ideal form, appears in Fig. 7 [4]. These plots are often called *Nyquist diagrams.* It is more common to plot the data as imaginary impedance versus real impedance. This method has been used to study the behavior of inhibitors, pitting attack on different alloys, organic coating behavior in solutions, and the corrosion of alloys in low-conductivity solutions.

Passivation

As described earlier in this chapter, when the anodic current density is sufficiently high precipitation of a corrosion film becomes likely. This is especially true for many metals exposed to corrosive environments. The anodes and cathodes are continually moving around the surface of the metal, so the double layer contains a combination of oxidation products and reduction products. For example, in

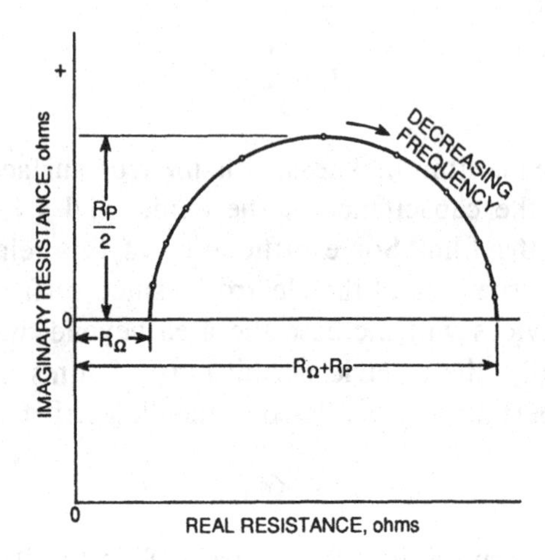

Figure 7. Real versus imaginary resistance values measured as a function of frequency.

the case of iron corroding in neutral pH water containing dissolved oxygen, the reactions are

$$2Fe \rightarrow 2Fe^{2+} + 4e \qquad \text{(oxidation)} \tag{15}$$

$$O_2 + 2H_2O + 4e \rightarrow 4OH^- \qquad \text{(reduction)} \tag{16}$$

with a net reaction of

$$2Fe + 2H_2O + O_2 \rightarrow 2Fe^{2+} + 4OH^- \rightarrow 2Fe(OH)_2 \tag{17}$$

the last product on the right of equation (17) precipitating as a solid. As long as corrosion is generally uniform, the precipitation of $Fe(OH)_2$ over the surface will be relatively uniform. Once the corrosion film is established, ferrous ions must diffuse out through this film to the film/solution interface to continue corrosion of the bulk metal, or oxygen and water must diffuse into the film to the metal/film interface. Transport across the film in either direction will reduce the rate of reaction, thus decreasing the corrosion rate. This simplistic model is the basic nature of passivation; and the theme of this book.

The thermodynamics of passivation have briefly been discussed in Chapter 1 and will be presented in detail in subsequent chapters. The remainder of this chapter will deal with the kinetic aspects of passivity, again, only briefly because the remainder of the book will be specific on kinetics in other regards.

Many years ago H. H. Uhlig defined two types of passivity [5]. The first is when an active metal in the EMF series becomes passive when its electrochemical behavior is similar to a noble metal in the EMF series; that is, the corrosion rate becomes low and the potential more noble. The second type of passivation is when a metal resists corrosion in an environment when there is a large thermodynamic driving force for the formation of a corrosion film.

Examples of the first type are the many metals and alloys that naturally passivate in air, such as chromium, nickel, molybdenum, titanium, zirconium, stainless steels, and iron in oxidizing environments. The second type of passivation is found for lead in sulfuric acid, magnesium in water, and iron in inhibited pickling acid. In these cases, low corrosion rates are observed, but the corrosion potential is still relatively active and, therefore, alloys with this behavior are not passive according to the first definition.

Figure 8 is an example of the first type of passivity. Anodic dissolution occurs at an ever-increasing rate until a critical current density, i_c, is reached, at which point a substantial drop in corrosion current occurs to the passive current density, i_p, at a specific potential, the passivation potential (E_{pp}). As potentials become more noble, the corrosion current remains low over the entire passive region. The onset of passivity is coincident with the formation of a stable passive film. The constant dissolution rate (i) with increasing potential can be attributed to either of two possible mechanisms. Dissolution in the passive region is occurring by the transport of ionic species through the film as a result of the electric field across the film. The increase in potential through the passive region is accompanied by thickening of the film such that the electric field across the oxide remains constant. The second possibility is that the current is controlled by the dissolution rate of the film, which is a chemical process and, therefore, potential-independent. That is, the current is just sufficient to replace the film that dissolves at the film/solution interface such that a dynamic equilibrium is attained.

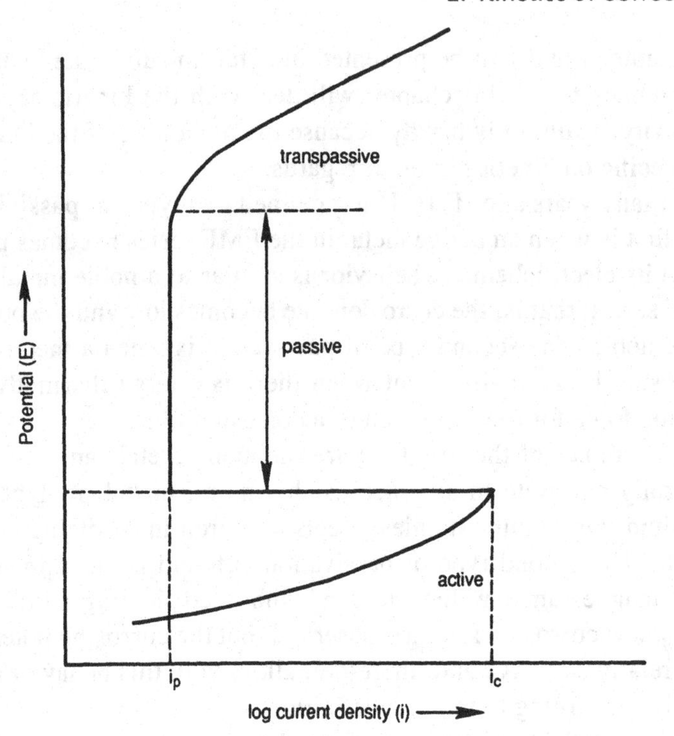

Figure 8. Ideal anodic polarization curve for a metal displaying classic passivation.

Figure 8 demonstrates that passivity or the protectiveness of a corrosion film is a kinetic phenomenon, and the thermodynamic aspects of passivity discussed in Chapter 1 are not sufficient to predict the degree of protection (corrosion inhibition) provided by a reaction film.

The ability of a corrosion film to offer corrosion inhibition to the substrate is a complex and interdependent function of many factors. Primary among these film factors are chemical composition, thickness, crystalline structure, electronic properties, including defect structure, and mechanical properties. These various factors will be discussed in greater detail in the following chapters.

There are two basic theories of passivity; one accounts for the metals that passivate according to the first definition of passivity, the other for the second definition of passivity [6]. The first, referred to as the *adsorption theory,* suggests that metals such as chromium, aluminum, titanium, etc., are covered by a monolayer or less of chemisorbed ions (i.e., O_2 on Ti). This layer of ions displaces adsorped water molecules and slows down the rate of metal dissolution. Since only a monolayer or less of adsorbing ions is present, the film cannot act as a diffusion barrier in the same manner as a thick film. These two-dimensional films are typically invisible to the unaided eye. The other theory, sometimes referred to as the *oxide film theory,* stipulates that a three-dimensional film forms, which acts as a diffusion barrier and thus retards further reaction of the substrate. Furthermore, the film has inherent electronic properties that are the principle cause of passivation. These films may be visible to the unaided eye and are usually 10 to 100 Å thick.

While these two mechanisms have generated much discussion and controversy over many years, the advent of surface science techniques to study monolayer coverage on surfaces has essentially resolved the matter. The passive film can be either two-dimensional, about one monolayer thick, or three-dimensional, depending on the metal–solution system. Moreover, a transition from one to the other can occur in the same system, depending on the potential. For example, Frankenthal [7] found at low potentials in the passive region a film less than the unit cell thickness for either Fe_3O_4 or γ-Fe_2O_3 was present on iron in dilute NaOH buffered with borate. This indicates the film was essentially adsorbed oxygen. At higher potentials, the film thickness exceeded the unit cell dimension for the ferric oxide γ-Fe_2O_3, implying a three-dimensional film. Other examples exist of this transition, from two-dimensional to three-dimensional films. Some of these examples will be discussed in later chapters.

The importance of electrochemical methods in providing information on the metal/solution interface, and, more specifically, the contribution of films to corrosion behavior cannot be overemphasized. The combination of techniques such as AC impedance and surface analysis methods provide the tools necessary to develop a complete understanding of the role and contribution of films on all forms of corrosion. When complete understanding of how films nucleate, grow, and interact with the environment and either enhance or inhibit the degradation of metals is achieved, methods to combat corrosion can be developed that will incorporate the properties of the films themselves.

High-Temperature Oxidation

Corrosion films formed at high temperature (nonaqueous) are often described as either scales (when oxide layers are thick) or tarnishes (when oxide layers are thin). The kinetics of tarnish and scale formation are initially determined by the reaction rate of the metal surface with the gaseous environment. Similar to aqueous corrosion, the process is electrochemical in nature. The anodic reaction occurs by oxidation at the metal–scale interface:

$$M \rightarrow M^{2+} + 2e \qquad (18)$$

The cathodic reaction occurs by reduction at the scale-gas interface:

$$\tfrac{1}{2}O_2 + 2e \rightarrow O^{2-} \qquad (19)$$

Both produce the net reaction

$$M + \tfrac{1}{2}O_2 \rightarrow MO \qquad (20)$$

The oxide grows at the metal–scale interface or at the scale–gas interface, depending on the rate-controlling process that will be discussed in Chapter 3.

The rate of scale growth thus depends on the transport of cations and anions in either direction across the scale to either interface. Since there is also a transfer of electrons, the electronic conductance is also an important factor during high-temperature oxidation. The scale is most often firmly attached to the base metal; thus high-temperature oxidation is generally expressed as a weight gain per unit area of surface.

Four primary oxidation rates have been observed for many metals in various environments. These have been designated as laws but are actually empirical relationships described as linear, logarithmic, parabolic, and cubic. The linear rate law is

$$\frac{dx}{dt} = k_L \qquad (21)$$

where x is the oxide thickness, t is the oxidation time, and k_L is the rate constant for linear oxidation kinetics. Equation (21) can be integrated to give

$$x = k_L t + C \quad (22)$$

where C is a constant of integration.

This is the simplest form of high-temperature attack, and illustrates that the rate of alloy destruction is unhampered by oxide formation and will proceed at a constant rate. The constant is a function of gas composition, pressure, temperature, and type of metal. Figure 9 shows the oxide mass formed as a function of time for linear kinetics. Linear rate behavior is demonstrated by scales that are cracked or porous, eliminating the barrier effect of the oxide. Sodium and potassium display linear kinetics.

Another oxidation rate law is logarithmic, and its sequel is the inverse logarithmic law. The results of oxidation that follow this behavior can be equally presented either way; therefore, the two cannot be distinguished. The logarithmic equation is

$$x = k_e \log(at + 1) \quad (23)$$

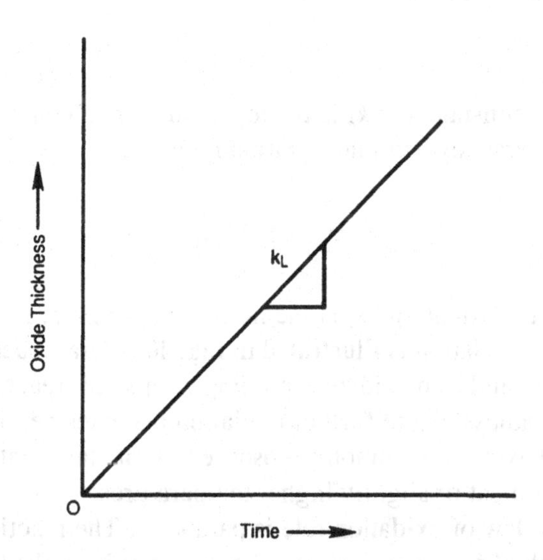

Figure 9. Linear oxidation kinetics for high-temperature scaling of a metal or alloy.

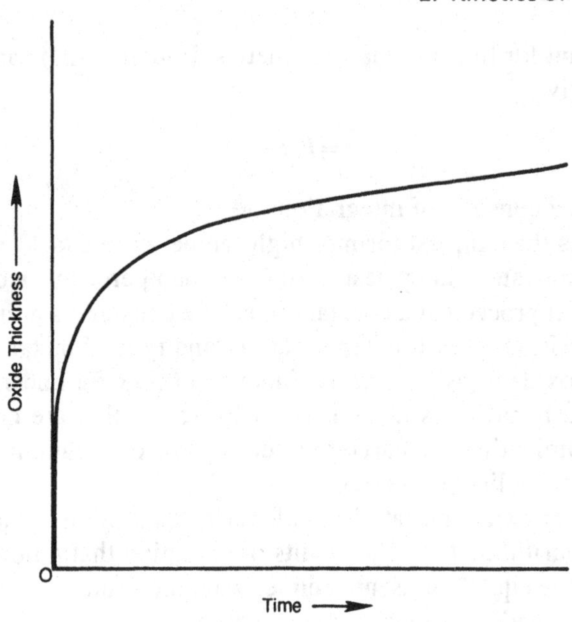

Figure 10. Logarithmic and inverse logarithmic oxidation behavior for high-temperature scaling.

where a is a constant and k_e is the logarithmic rate constant.

The inverse logarithmic equation is

$$1/x = b - k_i \log t \tag{24}$$

where b is a constant and k_i is the inverse logarithmic rate constant. This type of oxidation is illustrated in Fig. 10. Metals such as aluminum, copper, and iron oxidize according to this law, reaching a limiting film thickness where further oxidation becomes negligible. This type of behavior is commonly observed on metals that form thin oxides at ambient or slightly higher temperatures.

A third law of oxidation rate is parabolic. The reaction kinetics are the result of a chemical potential gradient of ions, the diffusion of which through the oxide is the rate-controlling step. The diffusion flux is inversely proportional to the oxide thickness:

$$\frac{dx}{dt} = \frac{k_p}{x}$$

where k_p is the parabolic rate constant. Upon integration,

$$x^2 = 2k_pt + C \quad (25)$$

This type of oxidation is often observed when thick coherent oxides form, typically at high temperatures where the mobility of ionic components is sufficient to develop such a scale. Metals such as Co, Cr, Fe, and Cu form these type of scales at high temperature. Figure 11 shows an example of parabolic kinetics.

The fourth type of oxidation law is referred to as cubic, although this law can also be described mathematically as an intermediate stage (transition) between logarithmic and parabolic kinetics. The cubic equation is

$$x^3 = k_ct + C$$

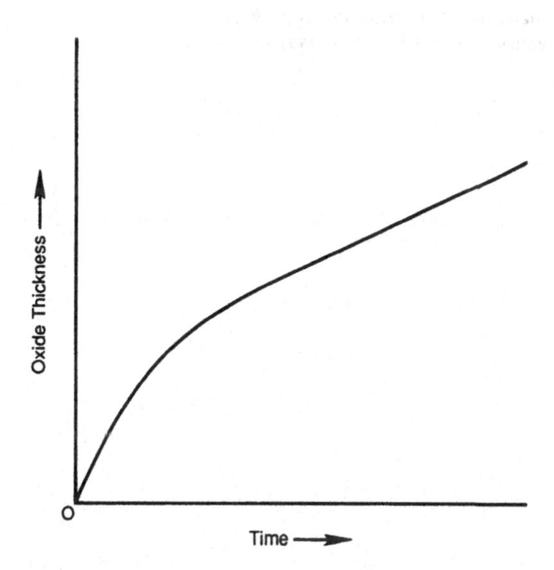

Figure 11. Parabolic reaction kinetics for high-temperature scaling.

where k_c is the cubic reaction kinetics constant. This curve would also be intermediate between the logarithmic (Fig. 10) and parabolic (Fig. 11) curves illustrated.

Comparison of Fig. 9 to 11 indicates that a linear oxidation rate is a catastrophic form of attack, while logarithmic and parabolic are more desirable for high temperature. These two forms of oxidation come closest to passivation behavior for high-temperature oxidation. Neither strictly meets Uhlig's definition for low-temperature passive behavior, but the corrosion rate is substantially reduced at high temperature once a stable protective film is formed.

References

1. J. O'M. Bockris and A. K. N. Reddy, *Modern Electrochemistry,* vols. 1 and 2, Plenum, New York, 1977.
2. D. W. Shoesmith, Corrosion, vol. 13, p. 29, *ASM Metals Handbook,* ASM International, 1987.
3. F. Mansfeld, *Corrosion* **1981**. *36,* 301.
4. S. W. Dean, Jr. Corrosion/85, Paper No. 76, NACE, Boston, MA, 1985.
5. H. H. Uhlig and R. Revie, *Corrosion and Corrosion Control,* 3rd ed., Wiley, New York, 1985.
6. J. Kruger, *International Materials Reviews,* **1988**, *3,* 113.
7. R. P. Frankenthal, *Electrochim. Acta,* **1971** *16,* 1845.

3

Properties of Corrosion Films

Introduction

This chapter presents the fundamental aspects of corrosion films and their properties. The stages of film formation and growth are discussed beginning with the adsorption of two-dimensional layers and the eventual development of thin and thick interphase films. Film growth kinetics are discussed along with the type of defects that develop and their impact on the properties of films. Finally, the stability of the films that form is discussed.

Many investigators differentiate high-temperature scales that form on metals under gaseous conditions from low-temperature films that form in aqueous solutions. Due to the many similarities and characteristics of both systems, we combine them here for ease of understanding. One very important similarity is the electrochemical process of degradation. Both high-temperature scaling and low-temperature corrosion are electrochemical phenomena [1]. The electrochemical aspects of high-temperature oxidation have been discussed briefly in Chapter 2.

When a specific attribute of one film is particularly important, the discussion will center on either high-temperature scaling or low-temperature oxidation. However, most often the two extremes have similar mechanics and will be treated as such.

Interface Reactions

Interface reactions are those initial reactions of a clean metal surface with an anion species by chemical adsorption. This adsorption requires no activation energy and, therefore, is a very rapid process. At $-195°C$ oxygen forms a stable monolayer on molybdenum and iron develops a multilayer oxide. In fact, this rapid adsorption has been found to occur even at liquid helium temperatures. Since real surfaces are not truly flat, the initial interaction of foreign atoms with the metal surface is best described by the *terrace-ledge-kink* (TLK) model [2]. Figure 1 is an ideal representation of a metal surface that contains terraces, ledges, and kinks. It is thermodynamically most favorable for an adatom (foreign atom) to bond at a kink rather than a ledge or a terrace. The particular TLK appearance of a surface depends on the crystal structure of the metal, the crystallographic orientation of the specific planes at the surface, and any misorientation of these planes.

As the temperature increases, the stability of metal surface atoms decreases until above a certain critical temperature the surface roughness becomes so great due to thermal effects that the terraces and ledges become indistinguishable. This critical temperature has been determined to be [3]

$$T_R = 0.63\epsilon/k \tag{1}$$

where ϵ is the binding energy between nearest neighbors and k is

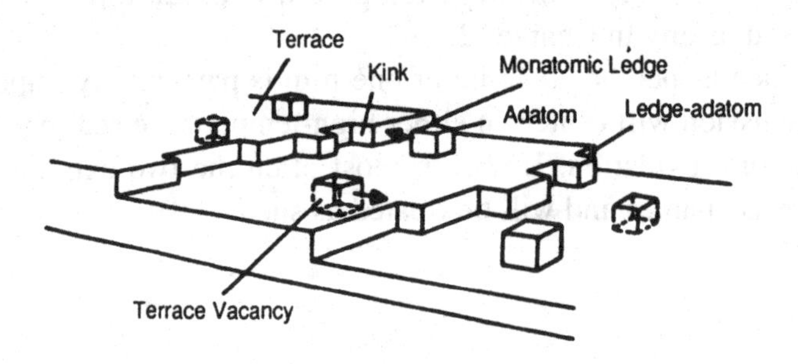

Figure 1. Terrace-ledge-kink (TLK) model of a surface.

Boltzmann's constant. Above this critical temperature, the TLK model breaks down.

The development of a monolayer or some fraction of a monolayer of adsorbate on the surface of a metal depends on numerous factors. Principal among these are the [2]

- ☐ Number of bond states
- ☐ Population of each of these states
- ☐ Bond energy associated with each state
- ☐ Electron transfer between metal and adsorbate
- ☐ Structure of the adsorbed layer
- ☐ Interactions between adsorbed atoms
- ☐ Kinetics of adsorption and desorption

The first four factors deal with the nature of bonding between the metal and the adsorbed atom. The next two are structural factors that depend on the specific adsorbed atoms and how they develop the monolayer. The last factor is related to the kinetics of adsorption/desorption, which determines whether a stable monolayer will develop or whether desorption will predominate, thereby limiting or eliminating oxidation. This reversibility of chemical adsorption is temperature-dependent and, for gases, partial pressure–dependent. Figure 2 is typical of a reversible adsorption isotherm for S on Cu [4]. Between 800 and 900°C the surface coverage is a function of the H_2S partial pressure. At 10^{-2} atm a complete monolayer of sulfur coverage is obtained; however, it is a reversible condition. No three-

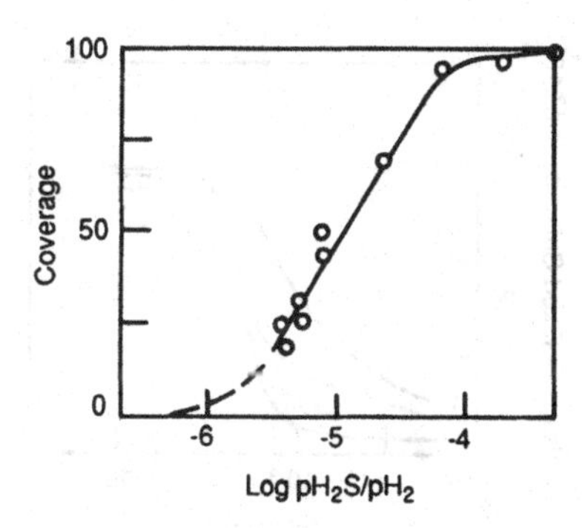

Figure 2. Adsorbed sulfur coverage on copper as a function of sulfur pressure.

dimensional, irreversible CuS has formed. Below about 10^{-6} atm no adsorption of sulfur on the metal surface occurs.

Besides temperature and partial pressure dependence of adsorption, there is a crystallographic dependence. This is illustrated in Fig. 3, again for Cu in H_2S [4]. Note that at high partial pressures of H_2S the crystallographic dependence is minimal, while at very low partial pressures there is a significant dependence.

Thus, it can be expected that at low partial pressures nucleation of the two-dimensional structure will occur at preferential crystallographic locations. In some instances the adsorption of foreign atoms on these crystallographic planes will fill energetically favorable lattice sites in the preexisting substrate matrix that maintains the preferred orientation of the substrate. This preferred orientation of the two-dimensional structure over the substrate is called *epitaxy*. Epitaxy is quite common in many metal-oxide systems.

The further development or growth of the two-dimensional film structure occurs by lateral coverage across the metal surface between so-called islands that have preferentially nucleated. The growth rate of this monolayer will depend on the temperature and partial pressure of the gas. At high partial pressures, the surfaces interacting with the gas are more numerous and, therefore, fast growth will occur. Likewise, with higher temperatures surface diffusion of the adsorbate will be stimulated, leading to enhanced growth.

A similar picture evolves for low-temperature aqueous adsorption. However, this process is complicated by such factors as the

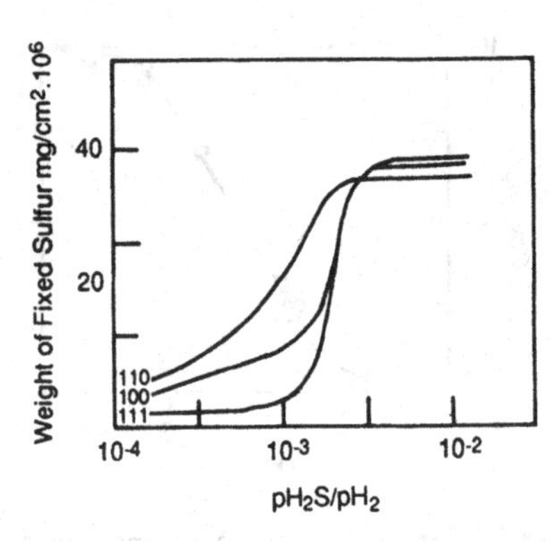

Figure 3. Crystallographic dependence of sulfur adsorption on copper.

electric double layer and adsorption of cations at the metal/solution interface. Once a complete monolayer has been established by coalescence of the islands, further growth of the oxide is generated by transport across the oxide film.

Interphase Development

Even before adsorption has covered the entire surface of the metal, the corrosion film may begin to grow beyond monolayer coverage and thickening begins to occur. The development of an interphase that exists between the metal and solution is discussed in the remainder of this chapter.

Corrosion films, as they will be described here, are also referred to as oxides, reaction products, scales, and tarnishes. The term corrosion film is used to emphasize the corrosion aspect of the film development rather than the term *oxide,* which misrepresents the role of sulfides or scales and may be confused with the precipitation of mineral scales (e.g., $BaSO_4$, $CaCO_3$) in aqueous solutions that have no bearing on the corrosion-generated films.

As was discussed in Chapter 1, thermodynamics is valuable for describing the overall driving force for corrosion and the generation of corrosion films. Pourbaix diagrams and free-energy–temperature diagrams (Richardson), as discussed in Chapter 1, are useful in providing data on phases that can be expected to form under specific temperatures and pressures. Moreover, at high temperatures isothermal stability diagrams have been utilized. These diagrams, referred to as *Kellogg diagrams,* are generated using standard free energies of formation for all elements and compounds expected to be present at a fixed temperature as a function of gas partial pressure. Figure 4 is an example of such a diagram for Ni exposed to O and S at 1250 K [5]. Thus, from this diagram the partial pressures of S and O necessary to form $NiSO_4$ are apparent, and the regions of predominance for NiS compared to NiO are shown. Also, it is evident that under certain partial pressures of S and O, Ni metal is stable with no formation of a stable corrosion film. Variations on this type of diagram include multielement, multigas systems (Fig. 5) [5].

Of course, these diagrams, as with Pourbaix diagrams, are lim-

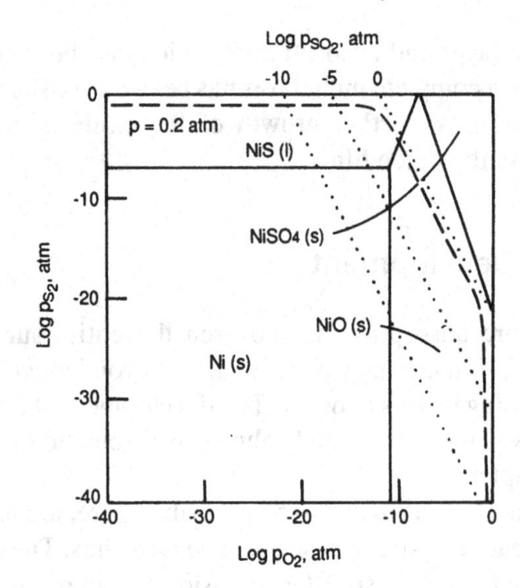

Figure 4. Phase stabilities in the Ni/O/S system at 1250 K.

ited because of the thermodynamic equilibrium imposed, which in practice is often not present, and the implicit assumption that the films are a uniform, continuous, defect-free interphase.

The growth of corrosion films on metal surfaces often involves preferential development on specific crystallographic planes, just as discussed for interface reactions. For example, the growth of TiO_2 on titanium depends on the index of the specific plane (Fig. 6) [6]. The basal plane [0001] is observed to form a thin oxide that has been found to be quite protective, whereas other low-index planes form thicker films that are less protective. Thus, the texture of a metal can have a significant effect on corrosion rate, at least in the early stages of film development.

Similar results have been observed for Zn, another hexagonal close packed (hcp) metal [7]. The basal plane [0001] had the most completely formed oxide when Zn was immersed in sodium hydroxide and correspondingly the lowest capacitance when compared to films formed on [1010] and [1120] planes.

The generalized macroscopic growth laws of corrosion films were presented in Chapter 2. However, it has been found that for crystallized corrosion films the product that forms is not stoi-

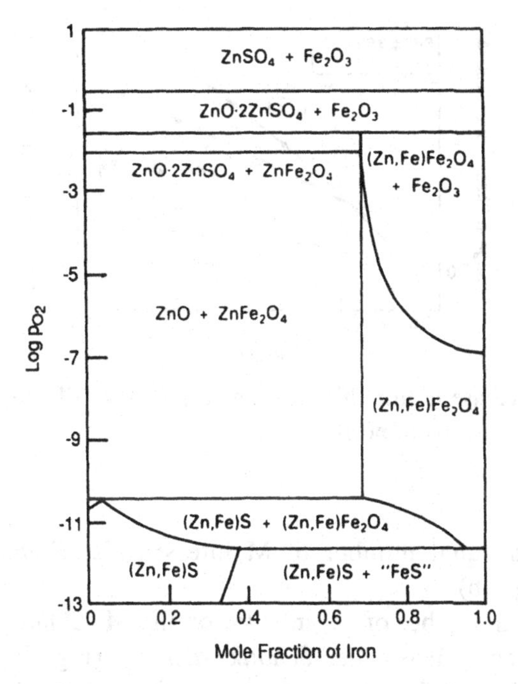

Figure 5. The Fe/Zn/S/O system at 1164 K and 1 atm sulfur dioxide gas pressure.

chiometric as implied from thermodynamics but may vary considerably from stoichiometry. This nonstoichiometry is the result of defects in the crystal lattice. The type, concentration, and distribution of these defects are important to the overall behavior of corrosion films and, therefore, the eventual corrosion of the metal substrate.

Defects in Crystalline Solids

In stoichiometric compounds there are five basic types of disorder. They are [8]

1. Equal number of vacancies on anion lattice sites (X) and the formation of X interstitial atoms with no perturbation of the M sublattice. This is called a *Frenkel disorder* (Fig. 7a).
2. Same as No. 1, but vacancies are on the cation sublattice (M)

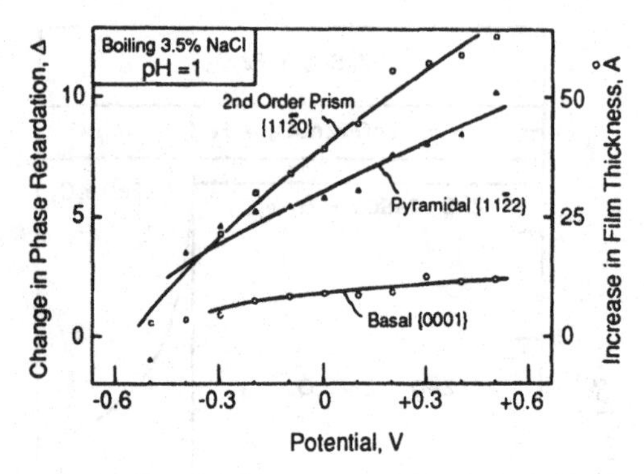

Figure 6. Increase in film thickness on various crystallographic planes of titanium as a function of potential.

with equal number of M interstitials: *Frenkel disorder* (Fig. 7b).

3. Equal number of vacant sites on the M sublattice and the X lattice. This is called *Schottky disorder* (Fig. 7c). The example in Fig. 7c is for anions and cations of equal valence.
4. Equal numbers of M and X interstitials (Fig. 7d).
5. Substitutional disorder between the M sublattice and the X lattice (Fig. 7e).

Since the compound must be electrically neutral, the combination of these different disorders must always be balanced to attain charge neutrality. In order to maintain this neutrality, electronic defects must also be accounted for in the defect structure. However, for truly stoichiometric crystals, the electronic defect concentration is generally negligible. In reality few corrosion films are stoichiometric; more often they are nonstoichiometric, and the five defect structures presented above do not necessarily have to occur. Only charge neutrality must be maintained as deviation from stoichiometry proceeds. Most oxide and sulfide corrosion films are nonstoichiometric. Ionic diffusion by Frenkel disorder, as in Fig. 7b, and Schottky defects are most likely to occur and may be the rate-controlling step for oxidation of most metals. The deviation from stoichiometry of most corrosion films has been found to produce semiconductive behavior. These

```
  M X M X M X M X            M X M X M X M X
            X                      M
  X M   M X M X M            X   X M X M X M
                                     M
  M X M X M   M X            M X M X M X   X
   X     X
a X M X M X M   M          b X M X M X M X M


  M X M X M X M X            M X M X M X M X
                                   M
  X   X M X M X M            X   X M X M X M

  M X M X M   M X            M X M   M X M X
                                  X
c X M X M X M X M          d X M X M X M X M


              M X M Ⓜ M X
              X M X M X M
              M X ⓧ X M X
            e X M X M M M
```

Figure 7. Common crystal disorders: (a) and (b) Frenkel, (c) Schottky, (d) equal numbers of interstitials, and (e) substitutional disorder.

semiconductor properties of corrosion films also control the electronic defects in the film. Moreover, some films behave as insulators.

Electrons in semiconductors fall into one of two regions: the valence band or the conduction band. Figure 8 illustrates the energy gap between the conduction band and the valence band of a compound. This energy difference between the two bands, or, more properly, the *band gap energy*, is a measure of the degree of

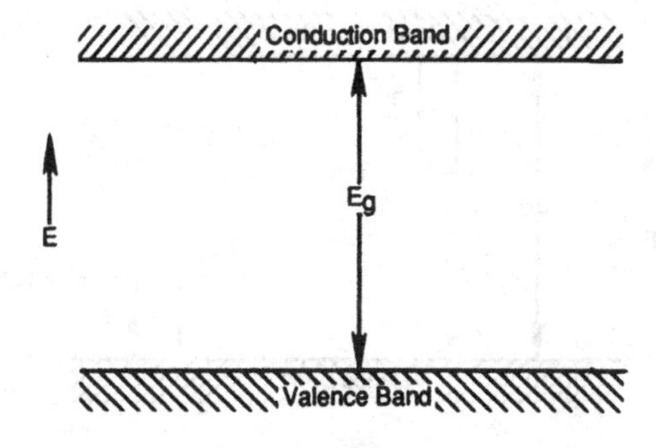

Figure 8. Ideal representation of the band gap between the conduction band of a compound and the valence band.

semiconductivity of the compound or, conversely, of the insulator properties of the compound. As the band gap energy increases, the compound behaves more like an insulator.

At absolute zero the valence band is completely filled and the conduction band is completely empty. Therefore, the compound is an ideal insulator—no current will flow. As the temperature increases, electrons will have sufficient energy to cross the forbidden region in the band gap to the conduction band, establishing current flow. This process will result in unoccupied states in the valence band, or, as they are often called, *holes*. Thus, current flow is maintained as electrons are promoted to fill holes in the valence band and other electrons jump to the conduction band. This process is analogous to atomic diffusion in conjunction with vacancies.

The introduction of defects to a perfect crystal results in the contribution of electrons to the conduction band by the defect introduced or the acceptance of electrons from the valence band. If the electron energy levels of the defect lie within the forbidden energy gap of the crystal, as they often do in the case of impurity atoms, this process of electron jumping occurs. In effect, the defect narrows the band gap. Figure 9 illustrates the donor and acceptor states of such defects. If the defect contributes (donates) electrons to the conduction band, it is referred to as an *n-type defect*. If the defect accepts electrons from the valence band, it is called a *p-type defect*.

Corrosion films take on *p*-type behavior or *n*-type behavior depending on the predominant defect in the film. Thus, nonstoichio-

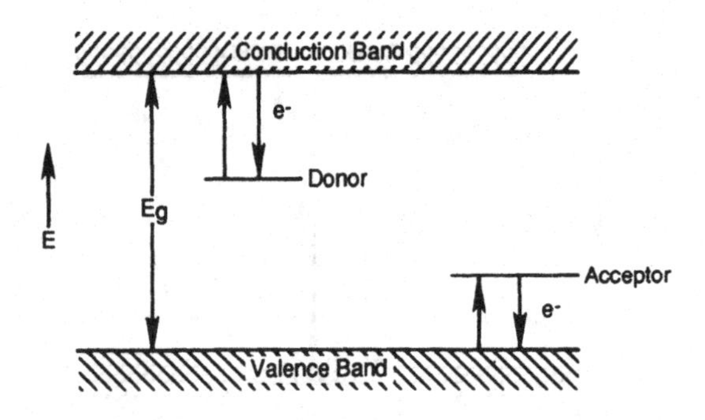

Figure 9. The effect of donor and acceptor defects on electron transport across the band gap.

metric corrosion films with cation vacancies will have *p*-type semiconductive properties since they accept electrons to maintain charge neutrality. To illustrate this, we write the following equation:

$$\tfrac{1}{2}O_2 \rightarrow Cu_2O + 2V_{Cu^+} + 2e^+ \tag{2}$$

where V_{Cu^+} represents a copper ion vacancy and e^+ is an electron hole. Thus, the generation of electron holes establishes the *p*-type behavior of the film.

Also, the mechanism of ionic transport in *p*-type oxides can be illustrated by Fig. 10. Here for cuprous oxide the cuprous ion diffuses in the direction of the film/solution interface by jumping into cation vacancies. This has the net effect of an opposite flux of vacancies toward the metal/film interface. Since ionic diffusion will be slower than electron flow, the former will be the rate-controlling step in the oxidation of copper.

In a similar situation, *n*-type behavior can be developed in corrosion films either by cation transport through interstitial diffusion or by anion diffusion inward toward the metal surface. Anion diffusion into the film proceeds via anion vacancy formation and anion jumping into these vacancies (Fig. 11).

For oxides the oxygen atom is typically many times larger than the metal atom, so anion diffusion is generally less energetically favorable than cation interstitial transport, especially at low tem-

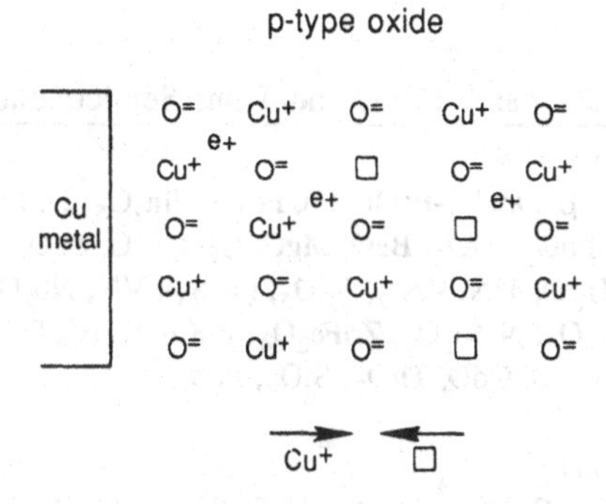

Figure 10. Schematic illustration of cuprous oxide as a *p*-type semiconductor. For every cation vacancy there is an additional electron hole for charge neutrality.

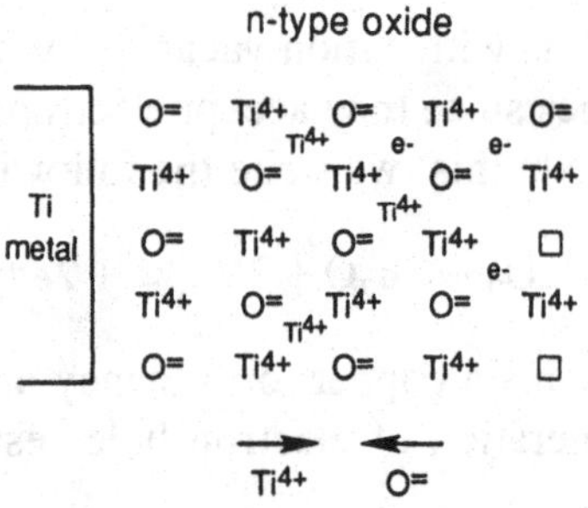

n-type oxide

Ti metal

O= Ti4+ O= Ti4+ O=
Ti4+ e- e-
Ti4+ O= Ti4+ O= Ti4+
Ti4+
O= Ti4+ O= Ti4+ □
e-
Ti4+ O= Ti4+ O= Ti4+
Ti4+
O= Ti4+ O= Ti4+ □

→ ←
Ti4+ O=

Figure 11. Schematic illustration of TiO_2 as an *n*-type semiconductor. For every interstitial cation there is a corresponding electron for charge neutrality. *n*-type semiconductors may also possess anion vacancies, shown here as □.

peratures. At elevated temperatures both anion and cation diffusion may occur.

Table 1 lists many corrosion films and whether they display *p*-type or *n*-type semiconduction. Care must be taken in generalizing about the *p*- or *n*-type characteristics of a film as presented in Table 1. Most of these data are for high-temperature films and *p*- or *n*-type behavior may change over an extended temperature range. For example, FeS is included to illustrate that the most common low-temperature corrosion product of iron with H_2S is mackinawite, which is a non-stoichiometric metal excess film (*n*-type). However, at higher tem-

Table 1. Corrosion Films and Their Semiconductor Type

n-*Type (metal excess)*

Al_2O_3, NiS, ZnO, α-, β-, and γ-FeO(OH), Fe_2O_3, Ta_2O_5, FeS (mackinawite), SnO_2, Ag_2S, MoO_3, PbO_2, TiO_2, BeO, MgO, CaO, SrO, BaO, BaS, ScN, CeO_2, UO_3, U_3O_8, TiS_2, (Ti_2S), TiN, ZrO_2, V_2O_5, (V_2S_3), VN, Nb_2O_5, (Cr_2S_3), WO_3, WAS, MnO_2, $MgFe_2$, O_4, $NiFe_2O_4$, $ZnFe_2O_4$, $ZnCoO_4$, ($CuFeS_2$) CdS, HgS(red), $MgAl_2O_4$, Tl_2O_3, (In_2O_3), CdO, ThO_2, SiO_2, PbS

p-*Type (metal deficient)*

FeO, CuS, SnS, Cr_2O_3, FeS (pyrrhotite), CoO, Bi_2O_3, MoO_2, UO_2, (VS), (CrS), $MgCr_2O_4$, $FeCr_2O_4$, $CoCr_2O_4$, $ZnCr_2O_4$, (WO_2), MoS_2, MnO, Mn_3O_4, Mn_2O_3, ReS_2, NiO, NiS, CoO, (Co_3O_4), PdO, Cu_2S, Ag_2O, $CoAl_2O_4$, $NiAl_2O_4$, (Tl_2O), Tl_2S, (GeO), (PbO), (Sb_2S_3), (Bi_2S_3), Cu_2O)

peratures or under different aqueous conditions, the pyrrhotite form of FeS may develop, which is metal deficient. Many authors do not consider these differences, and FeS is almost always reported, often incorrectly, as a *p*-type film. Other compounds display similar behavior.

Mott and Schottky developed an equation relating capacitance of the corrosion film to potential, assuming the films were semiconductors. If semiconductivity is present, plots of their equation will give a straight line for $1/C^2$ versus V. Figure 12 shows a plot that obeys this behavior [9]. The slope of the line can be used to estimate the concentration of donors in the film, and the intercept represents the flatband potential.

The Mott-Schottky equation that relates the space-charge capacitance C_{sc} to the potential, V_m, is [10]

$$C_{sc}^{-2} = (2/eN_D\epsilon\epsilon_0)(V_m - V_{FB} - kT/e_0) \tag{3}$$

where e is the charge of an electron, ϵ_0 is the permittivity of free space, ϵ is the permittivity of the film, N_D is the donor density, V_m is the electrode potential, and V_{FB} is the flatband potential.

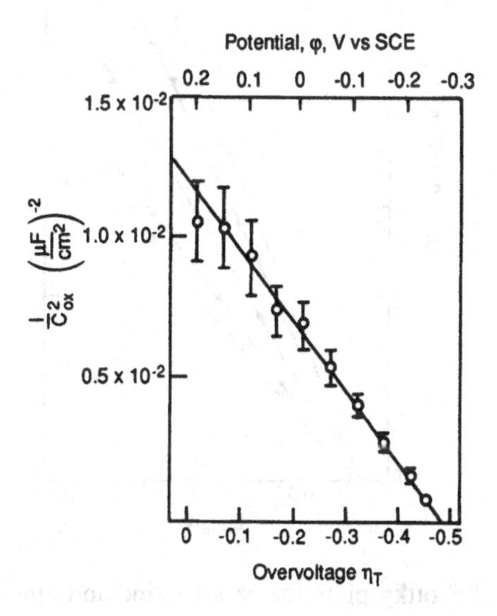

Figure 12. Mott-Schottky plot for passive iron at pH 8.4.

Figure 13 shows the effect of adding Co to Zn on the change in donor concentration and flatband potential [10]. Table 2 presents these changes, where it is apparent that increasing the Co content increases the concentration of donors in the ZnO corrosion film [11]. ZnO behaves as an *n*-type semiconductor. As discussed later, *n*-type films are often more resistant to uniform dissolution, but more susceptible to pitting corrosion.

In semiconductors, the addition of specific elements to either enhance or reduce conductivity is called *doping*. For *n*-type semiconductors the addition of lower-valent metal ions will reduce conductivity by decreasing the number of excess electrons and increasing the concentration of interstitial cations. Conversely, the introduction of higher-valent metallic ions will increase conductivity. For example, in ZnO, Zn has a valence of 2, so the alloying of Li, which has a valence of 1, into the oxide will decrease conductivity because the concentration of free electrons is decreased (Li will be an electron acceptor) for

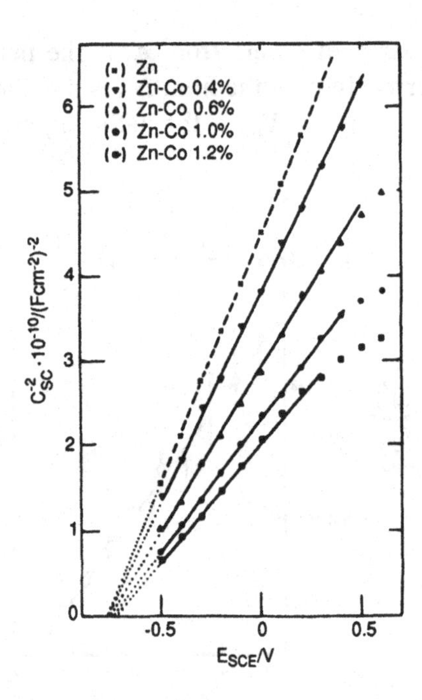

Figure 13. Mott-Schottky plots for passive zinc and zinc–cobalt alloys in 0.1 *M* NaOH solution.

Table 2. Donor Concentrations and Flatband Potentials of Passive Layers on Zn and Zn/Co Substrates

Substrate	$N_D/10^{20}$ cm^{-3}	E_{FB}/mV (SCE)
Zn	2.8	−770
Zn–0.4% Co	3.3	−760
Zn–0.6% Co	4.1	−760
Zn–1.0% Co	5.1	−735
Zn–1.2% Co	5.7	−725

every Li ion that substitutes for Zn^{2+}. This effect can be written in equation form as

$$Li_2O(\text{in ZnO}) + 2e_c + \tfrac{1}{2}O_2 = 2Li_{(Zn)'} + 2ZnO \qquad (4)$$

Where e_c represents excess conduction electrons; $Li_{(Zn)'}$ is for a Li atom substituted on a Zn atomic site with an apparent negative change. Thus, two conduction electrons are expended in the reaction and conductivity drops. Conversely, the addition of Al^{3+} to the ZnO will increase conductivity by increasing the concentration of conduction electrons:

$$Zn_{\circ\cdot} + Al_2O_3 = 2Al_{\bullet\cdot(Zn)} + e_c + 3ZnO \qquad (5)$$

where $\circ\cdot$ represents an interstitial lattice position of charge "dot," and $\bullet\cdot$ is an interstitial positive that has been filled by Al replacing the Zn atom. Therefore, for corrosion that is diffusion-controlled by vacancies or interstitials through the corrosion film, the corrosion rate will increase with the introduction of lower-valent cations and will decrease with the doping of higher-valent cations.

The same type of behavior, but in reverse, is observed for *p*-type semiconductors; that is, the introduction of lower-valent cations into a *p*-type oxide will increase the electron hole concentration and decrease the cation vacancy concentration, thereby decreasing the corrosion rate under diffusion control. Higher-valent cations will increase the vacancy content of the film and decrease the number of electron holes, thereby increasing the corrosion rate of a metal whose oxidation process is diffusion-controlled. Diffusion control is generally the rate-determining step in corrosion films, since electron and hole mobilities are orders of magnitude faster than ionic mobility.

The solution itself may also contribute to the donor density of the corrosion film. Figure 14 shows the donor density calculated from the slope of Mott-Schottky plots as a function of formation potential for iron in neutral borate and phosphate solutions [12]. At equivalent formation potentials the passive film on iron generated in phosphate solution has a greater defect concentration than for films formed in borate. Thus, the anions present in solution can have an effect on the number and type of defects produced in a growing corrosion film.

Generally, as the film continues to grow and age the defect density decreases [12] (Fig. 15), while the flatband potential remains essentially constant. This is consistent with observations of numerous investigators studying the contribution of films to stress corrosion cracking, pitting, inhibition, etc., that film aging most often decreases the corrosion rate.

The distinction between *n*-type and *p*-type semiconductive films can be related under certain circumstances to the sign of the exponent of the partial pressure of the corrodent. For example, at high temperature, a *p*-type oxide such as cuprous oxide reacts as follows:

$$\tfrac{1}{2}O_2 \rightarrow Cu_2O + 2V_{Cu^+} + 2e^+ \qquad (6)$$

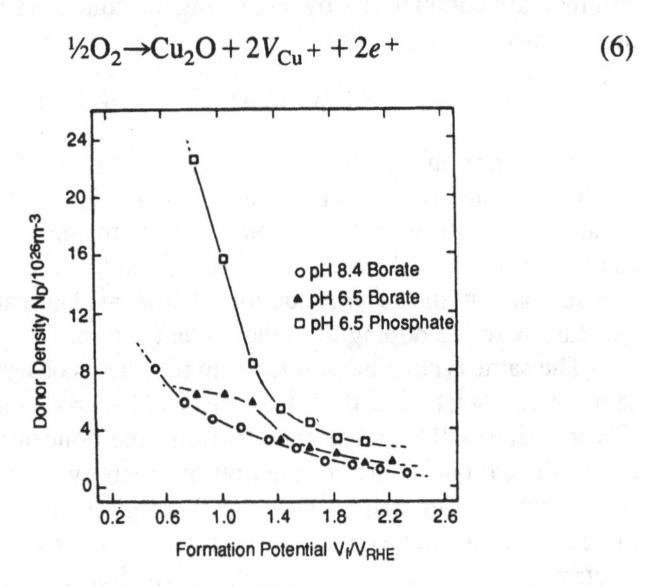

Figure 14. Donor density of the passive film formed on iron in borate (pH 8.4), borate (pH 6.5), and phosphate (pH 6.5) solutions as a function of film formation potential.

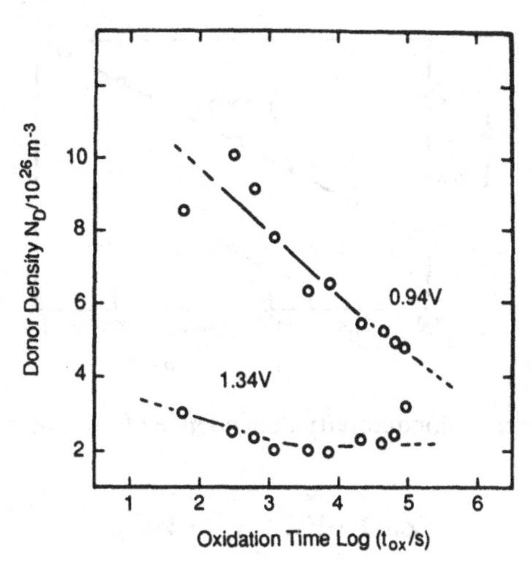

Figure 15. Donor density as a function of oxidation time for the passive film on iron at two potentials in borate (pH 6.5) solution.

where the electron holes (e^+) are the primary means of conduction. By the law of mass action,

$$K = \frac{(a_{Cu_2O})(a_{V_{Cu}^+})^2 a_{e^+}^2}{a_{O_2}^{1/2}} \tag{7}$$

where K is the equilibrium constant for the reaction and a is the activity of each component. Assuming a dilute solution of defects and that the partial pressure of oxygen is approximately equivalent to the activity, this equation reduces to

$$K = \frac{n^4}{p^{1/2}} \quad \text{or} \quad n = p^{1/8}K^{1/4} \tag{8}$$

where n is the concentration of defects, both ionic and electronic.

Here the concentration of defects is proportional to one-eighth the partial pressure of oxygen. This relationship is illustrated for NiO, which has a $p_{O_2}^{1/6}$ dependence (Fig. 16) on conductivity (defect concentration) [13]. Conversely, n-type films display an inverse relationship on oxygen partial pressure. For example,

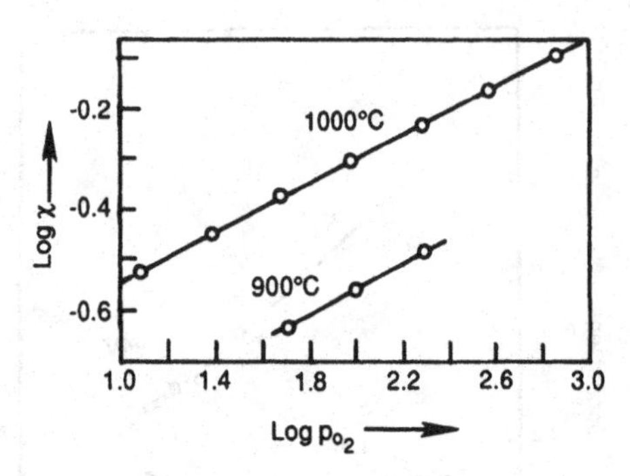

Figure 16. Electrical conductivity dependence of NiO on oxygen pressure.

$$ZnO \rightarrow V_{Zn^+} + e + \tfrac{1}{2}O_2 \tag{9}$$

and

$$n = \frac{K^{1/2}}{p_{O_2}^{1/4}} \tag{10}$$

However, for ZnO another reaction is possible:

$$ZnO \rightarrow V_{Zn^{2+}} + 2e + \tfrac{1}{2}O_2 \tag{11}$$

Then

$$n = K^{1/3} p_{O_2}^{-1/6} \tag{12}$$

Figure 17 demonstrates the defect dependence on oxygen pressure but with a slope intermediate between the two slopes predicted in the above equations. Thus, conduction may be from a combination of the two defect structures presented in equations (9) and (11).

The combined metal/film/solution system will develop a potential and charge distribution that is modified from that described for a metal in solution in Chapter 1. Figure 18 is a schematic representation of the charge distribution and potential change across the film and into the solution [14]. The charge distribution in the film is referred to as the *space-charge region* or *space-charge layer.* Equilibrium at the film/solution interface will be attained through a redistribution of electrons and holes in moving the space-charge region of the film and

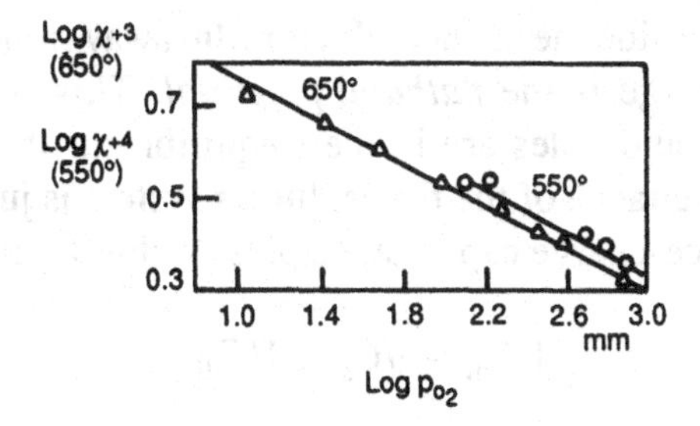

Figure 17. Electrical conductivity dependence of ZnO on oxygen pressure.

ions and dipoles in the double-layer region of the solution. Thus, a potential drop exists across the space-charge region that will have an effect on electrons and holes in the field. This will cause the energy bands (conduction and valence bands) to bend either up or down near the surface of the film, depending on the prevailing sign of the OHP. The energy bands bend down (potential drop) for a p-type corrosion film and bend up (potential rise) for an n-type semiconductor. The

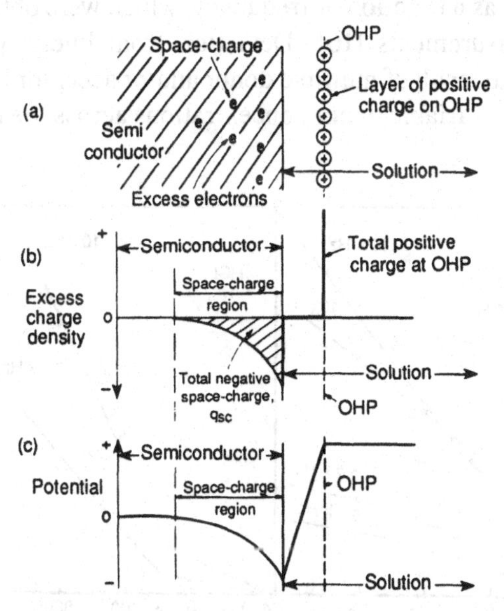

Figure 18. Schematic illustration of the (a) space-charge region inside a semiconductor, (b) the corresponding variation in charge density, and (c) the potential distribution.

electron energy inside the semiconductor film away from this bending by the space charge is the *flatband potential.* This is the potential where electrons and holes are in their equilibrium states. The total differential capacitance of the film/solution system is just the electrical series of space-charge capacitance plus double-layer capacitance:

$$1/C_{dc} = 1/C_{sc} + 1/C_{dl} \quad (13)$$

However, the double-layer capacitance is typically much greater than the space-charge capacitance, so the space-charge capacitance can be approximated by the differential capacitance. The space-charge capacitance and flatband potential have already been discussed in relation to the donor density of a film (equation (3)).

Mott-Schottky behavior is frequently observed and has been reported for oxidized Ti and Nb and CdS on Cd [15]. However, in many cases deviation from linearity is often observed. Figure 19 demonstrates the change in slope for oxides formed on titanium in different electrolytes and as a function of frequency, which were obtained from impedance measurements [16]. Deviation from linearity has been interpreted as the result of multiple donor and/or acceptor levels in the corrosion film or variable donor concentrations across the film. Suffi-

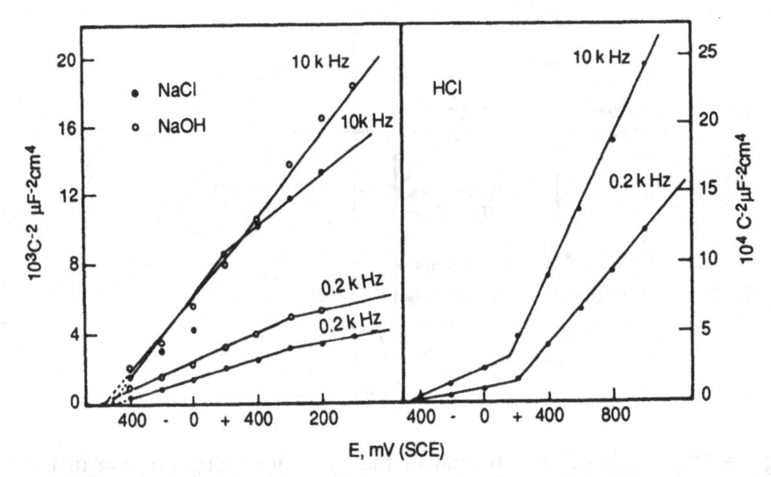

Figure 19. Mott-Schottky plots for the oxides formed at different potentials in 1 *N* NaOH, NaCl, and HCl.

cient data are not available to determine the cause of these variations in slope.

The space-charge region should not be confused with the surface charge of the corrosion film. A corrosion film will have a specific surface charge in a particular environment, depending on the pH. The pH for which there is zero net surface charge is referred to as the *pH of zero charge* (pH_{zc}), *point of zero charge* (pzc), or *isoelectric point* (IEP). The corresponding potential for the pH_{zc} is the *potential of zero charge* (E_{zc}). Thus, the charge on a corrosion film changes sign as the potential passes through the potential of zero charge. The pH_{zc} and E_{zc} are important to the study of corrosion films and eventually, therefore, to the corrosion of metals since these parameters determine the adsorption behavior of the film, which impacts the role of inhibitors and aggressive ions. The latter factor is important to the resistance or susceptibility of metals to pitting attack and stress corrosion cracking.

At pH values lower than the pH_{zc}, the surface of a corrosion film will have a net positive charge; at pH values higher than pH_{zc}, the surface charge is negative. The former condition will enhance the electrostatic attraction of anions such as halides, while the latter will enhance the attraction of cations. Since anions are most often associated with susceptibility to pitting and stress corrosion cracking of metals, a pH more acidic than pH_{zc} will promote such behavior. Some examples of the pH_{zc} are presented in Table 3 [15,17].

An empirical formula for oxides is [17]

$$pH_{zc} = 18.6 - 11.5z/d \tag{14}$$

where z is the oxidation state of the cation in the film, and d is the distance in angstroms of an adsorbed proton from the cation via the oxygen ion.

It has long been suspected that corrosion films generated in the dark would have different properties than those exposed to light. Recent evolution of photoelectrochemical techniques have confirmed this suspicion and have proven a useful tool for studying the semiconductive properties of corrosion films. Photoilluminescence of a corrosion film with a light beam of photon energy, $h\nu$, which is greater than or equal to the band gap energy, will excite electrons from the

Table 3. pH of Zero Charge for Various Oxides and Hydroxides

Material	pH_{zc}
α-Al_2O_3	5–9.2
γ-Al_2O_3	8.0
γ-$Al(OH)_3$	9.25
α-$Al(OH)_3$	5.0
CuO	9.5
Fe_2O_3	8.6
Fe_3O_4	6.5
CdO	12
Ta_2O_5	2.9
ZnO	8.8
TiO_2	4.7–6.2
$Pb(OH)_2$	11.0
$Mg(OH)_2$	12
SiO_2	2.2
ZrO_2	6.7
SnO_2	4.3
WO_3	0.4

valence band to the conductive band, producing a flow of current called the *photocurrent* [18] (Fig. 20). For an *n*-type corrosion film at potentials more positive than the flatband potential, the adsorption of photons creates electron/hole pairs that are separated by the electric field of the film. The electric field, referred to as a Schottky barrier, results from the mismatch in the Fermi level and the applied potential between the film and the solution. The electrons flow to the metal/film interface, the holes toward the film/solution interface. The holes react with anions, such as oxygen, producing a photocurrent. The magnitude of the photocurrent depends on the strength of the electric field.

It has been demonstrated that large band gap energy corrosion films such as ZnO dissolved anodically by consumption of holes, which for *n*-type semiconductors would only occur electrochemically under illumination [18]. Therefore, dark dissolution of ZnO is strictly a chemical reaction as follows:

$$ZnO + 2OH^- + H_2O \rightarrow [Zn(OH)_4]^{2-} \tag{15}$$

The dissolution rate of ZnO as a function of illumination is illustrated in Fig. 21 [19]. Other *n*-type oxides demonstrate the same

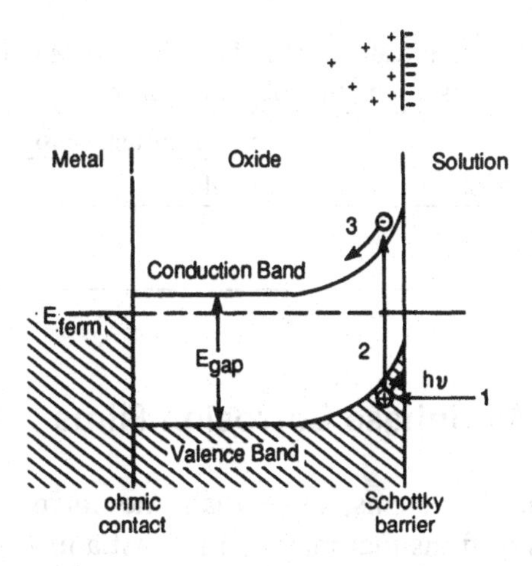

Figure 20. Schematic of the illumination of a semiconductor/electrolyte system. The photon strikes an atom in the valence band of the oxide (step 1), causing a hole/electron pair to form (step 2). The Schottky barrier moves the electron to the bulk film and the hole to the film/solution interface (step 3).

increase in dissolution rate and subsequent increased corrosion of the metal with illumination. For example, iron immersed in 0.01 *M* citric acid demonstrated significantly greater corrosion rates when illuminated than when in the dark [20] (Table 4). The corrosion films that initially form on steel in citrate are *n*-type semiconductors.

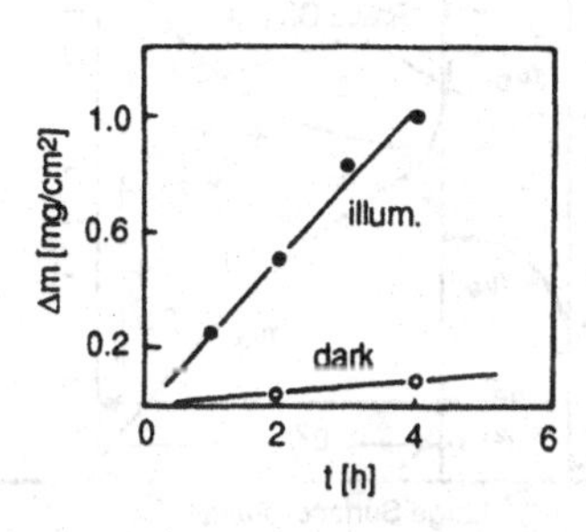

Figure 21. Rate of dissolution of ZnO as a function of illumination at 25°C and pH 4.6.

Table 4. Corrosion Rate of Mild Steel in 0.01 *M* Citric Acid as a Function of Illumination

	Corrosion current density (μA/cm²)	
Preimmersion time	Light	Dark
15 hours	78	49
27 days	190	130

Duplex and Multiphase Corrosion Films

In many metal systems, more than one corrosion film forms creating a series of films that maintain at least a metastable equilibrium across the various phase boundaries. Figure 22 illustrates this situation for iron in nitric acid [21]. The inverse spinel Fe_3O_4 develops adjacent to the metal surface and the ferric oxide Fe_2O_3 by contact with the solution. The space-charge region for each oxide is also presented. Oxygen ions migrate through the Fe_2O_3 film in response to the electric field, while the transport of iron ions is in the opposite direction and in competition with the oxygen ions. At the phase boundary between Fe_3O_4 and Fe_2O_3, there is an abrupt change in the concentration of electrons, n_e, and electron holes, n_{e^+}.

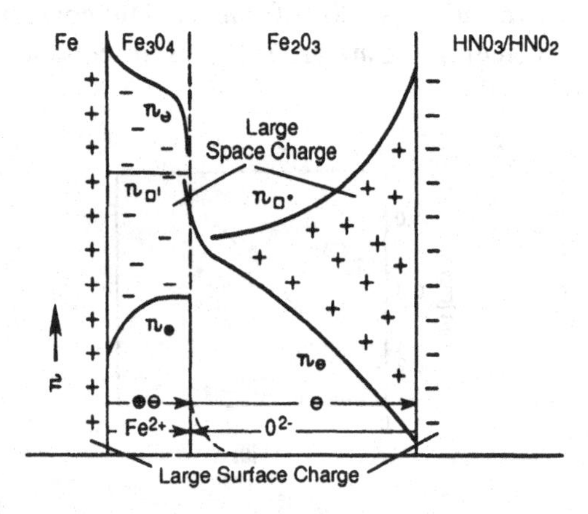

Figure 22. Illustration of the concentration gradients of electrons, holes, and ions in a dual-phase passive layer on iron.

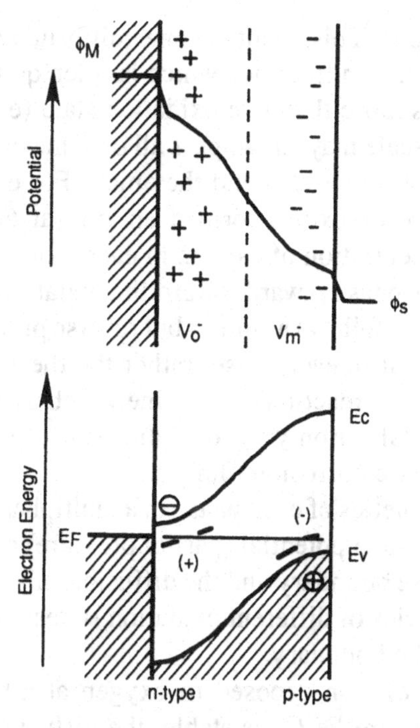

Figure 23. Idealized ion-selective bipolar passive film. The inner layer adjacent to the metal is an anion-selective, donor-rich, *n*-type layer, while the outer layer is cation-selective, acceptor-rich, *p*-type. V_O is an oxygen ion vacancy, V_M is a metal ion vacancy. The plus and minus signs represent positive holes and electrons, respectively. The plus and minus in parentheses signs represent ionized donors and acceptors, respectively.

If one film is a *p*-type semiconductor and the other an *n*-type, then a *p-n* junction as described by semiconductor physics may occur. Figure 23 illustrates this ideal behavior for a duplex corrosion film [22]. The *n*-type semiconductor film is adjacent to the metal surface and has excess electrons. Moreover, there are either anion vacancies or excess cations in this region. The outer film in contact with the solution is a *p*-type film with either cation vacancies or excess anions and holes.

At the junction, opposing diffusion of holes and electrons will initially occur. However, the electric field of the duplex film will oppose further diffusion of holes across the junction but promote the diffusion of electrons. This results in an equilibrium potential difference across the film/film interface which is often quite high.

If a metal has more than one oxidation state (e.g., ferrous, ferric), the resulting scale may encompass all or some of the compounds formed by these various states and the anion. For example, Fig. 24 shows the protocol of oxides formed an iron at 625°C. [22]. As expected, the concentration of oxygen in each oxide decreases when going from the gas phase inward toward the metal surface, while the concentration of iron follows a similar but reverse profile. Every oxide of iron is not present in every case; rather the thermodynamics and kinetics of the specific reactions determine which oxides will prevail. Conversely, if a metal has only one oxidation state, the resulting oxide will be a single-phase corrosion film.

The growth kinetics of each phase in a multiphase corrosion film depend on the chemical potential (partial pressure) of oxygen (for an oxide) at each phase boundary and the diffusion/migration kinetics of various cation species of different oxidation states through the phase on either side of the boundary.

For the case of iron exposed to oxygen at a temperature and oxygen pressure where Fe_2O_3 is stable, the diffusion processes and phase boundary reactions shown in Fig. 25 have been proposed [13]. The growth or decomposition of each phase is kinetically controlled and may follow any of the growth laws discussed in Chapter 2. Growth of Fe_2O_3 is principally by anion diffusion, whereas FeO is by cation diffusion. Fe_3O_4 growth is by both cation and anion diffusion. Below 570°C, FeO disappears from this series of reaction products.

It is well known that alloying iron with Cr increases the corrosion resistance of alloys both in aqueous environments and under high-

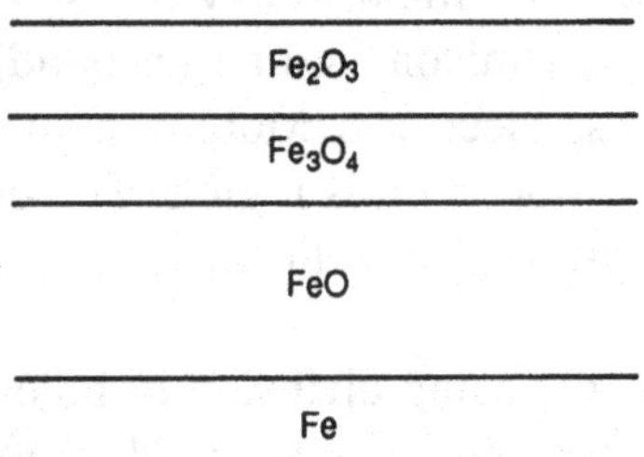

Figure 24. Protocol of oxides formed on iron at 625°C.

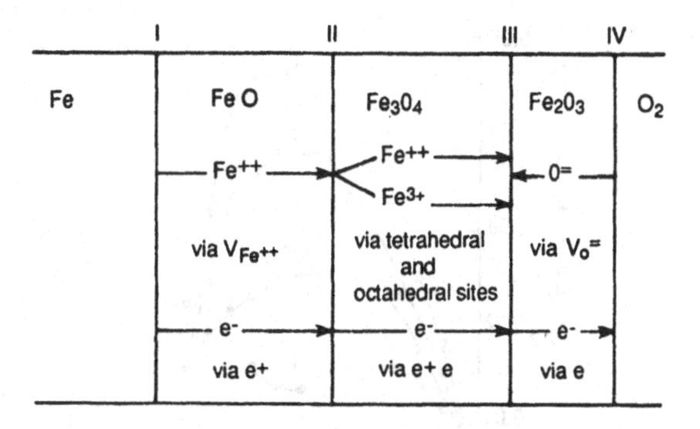

Figure 25. Schematic representation of the oxidation of iron in oxygen above 600°C. Note the following phase boundary reactions:
(I) $Fe + [V_{Fe^{2+}} + 2e^{+}]_{FeO} = 0$;
(II) $Fe_3O_4 = [4FeO + V_{Fe^{2+}} + 2e^{+}]_{FeO}[e^{+}]_{FeO} + [V_{Fe^{3+}}]_{Fe_3O_4} = 2[Fe^{2+}]_{FeO}$;
(III) $12Fe_2O_3 = 9Fe_3O_4 + [V_{Fe^{2+}} + 2V_{Fe^{3+}} + 8e^{+}]_{Fe_3O_4}$;
(IV) $\frac{1}{2}O_2(g) + [V_{O^{2-}} + 2e]Fe_2O_3 = 0$.

temperature oxidation. However, considering that FeO is a *p*-type semiconductor, film alloying with higher-valent Cr^{3+} should increase the oxidation rate, according to the earlier discussion on doping, rather than decrease it. Thus, the introduction of Cr in this case must modify the oxide in other ways to provide improved oxidation resistance. This modification may occur simply by altering the oxide from FeO to the spinel $FeCr_2O_3$ when sufficiently large concentrations of Cr are added (> about 5% Cr). At small concentrations of Cr (<1–2%), truly doped conditions, the corrosion rate may be accelerated as expected.

Deviation of Films From Semiconductor Behavior

Although much data exist to support thc semiconductor properties of corrosion films, sufficient data also exist to question the applicability of semiconductor concepts to all corrosion films under all conditions. For example, Fig. 26 illustrates the large deviation from linearity for Mott-Schottky plots of passivated iron as a function of film thickness [23]. The deviation at high anodic potentials indicates

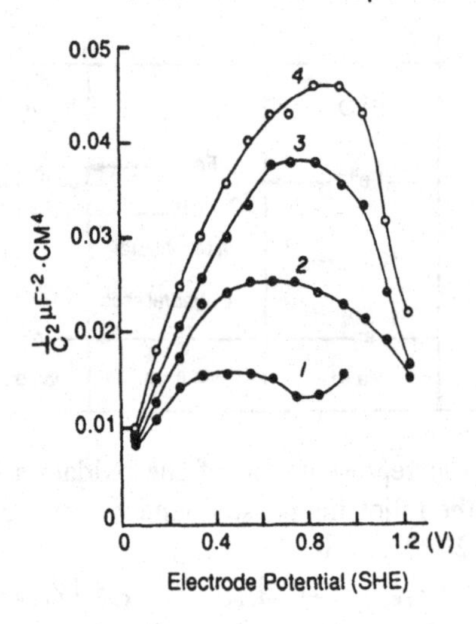

Figure 26. Mott-Schottky plots for passive iron at polarization time of 300 sec and different film thicknesses. Curve (1) 17 Å, (2) 20 Å, (3) 24 Å, (4) 27 Å.

an increase in capacitance after initial decreases in capacitance upon film formation and early growth. These deviations have been explained variously as resulting from slow states or deep donors in the films. However, another explanation has been put forward that contradicts the universality of classical semiconductor behavior of corrosion films [24]. These deviations can be explained by assuming the film is an insulator, the stoichiometry of which can be altered by valence state changes through oxidation and/or reduction reactions at the film/solution interface. Thus, for example, the ferric oxyhydroxide (FeOOH) formed on passive iron is considered an insulator with small nonstoichiometries of Fe^{2+} adjacent to the metal/film interface and "Fe^{4+}" ions near the film/solution interface. Fe^{4+} is considered to form at potentials where oxygen evolution occurs, induced by the local removal of protons. These disparities in ionic concentration create local film conductivity and produce defects that simulate semiconductor behavior. This behavior was given the name *chemiconductor* to distinguish it from classic semiconductor behavior of films. The general applicability of this model remains to be tested.

Discontinuous and Polycrystalline Films

The discussion thus far has centered on continuous, homogeneous films. However, corrosion films, especially thicker films often display a high degree of heterogeneity and discontinuity. For these cases, the semiconducting or defect aspects of the film may not be controlling or even an important factor in the further role of the film in corrosion of the substrate. Moreover, crystalline films are typically polycrystalline; therefore, an important contribution to further film growth and corrosion is the effect of grain boundaries on ionic diffusion. At high temperatures, corrosion film growth frequently follows a parabolic rate, first proposed by Wagner. Figure 27 shows this correlation for the oxidation of Ni above about 1000°C [25]. At this temperature, the oxidation of nickel was assumed to occur generally by cation diffusion in the oxide with electrochemical oxidation of the

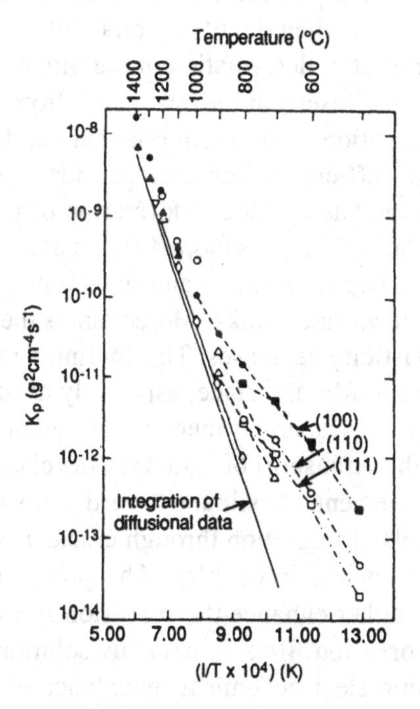

Figure 27. Parabolic oxidation rate constant for the oxidation of Ni to form NiO in 1 atm oxygen.

metal at the metal/film interface and reduction of the oxidant at the film/oxidant interface. However, as demonstrated in Fig. 27, below about 1000°C, in an intermediate-temperature range, accelerated corrosion occurs that is not consistent with the model. The increased scaling rate at these more moderate temperatures is expected to be due to the enhanced short-circuit diffusion of cations along the corrosion film grain boundaries. In fact, scaling calculations based on grain boundary diffusion rates for Ni in NiO in the range 500–800°C agree with this mechanism [26]. Thus, grain boundaries can have a significant effect on the protectiveness of films and the rate at which corrosion occurs. Additionally, scale dissolution or dissociation may occur preferentially along grain boundaries, creating fissures that will link the metal substrate with the environment or link the environment with a more porous inner film that is less protective.

The generation of porous films and the general breakdown of passivity at high temperatures is illustrated schematically in Fig. 28 [27]. In part (a) when the film is relatively new and thus thin enough to be quite plastic, deformation of the film occurs by dislocation glide and vacancy insertion into the metal from the film in conjunction with annihilation at the grain boundaries, dislocations, and second phases. At sufficiently high temperatures, grain boundary sliding also accommodates plastic deformation of the films. In part (b) vacancy annihilation becomes more difficult as a result of decreasing vacancy sinks. Dislocation climb and stabilization of the subgrain cell structure reduce vacancy sinks. Moreover, as the scale increases in thickness, its plasticity decreases. The continuing arrival of cation vacancies at the metal/film interface, especially at edges or corners, will generate voids. As scales continue to grow, an inner porous scale may form that inhibits transport of cations in the classical fashion but accelerates the movement of oxidant toward the substrate by providing easy channels of migration through cracks in the outer grains and voids or pores in the inner film. The presence of voids in a corrosion film can either enhance the corrosion rate or decrease it.

For porous corrosion films in aqueous solutions, a model has been proposed using electrochemical impedance to explain the observed behavior [28]. The frequent observation of depressed semicircles in Nyquist plots (see Fig. 7, Chapter 2) can be related to a charge

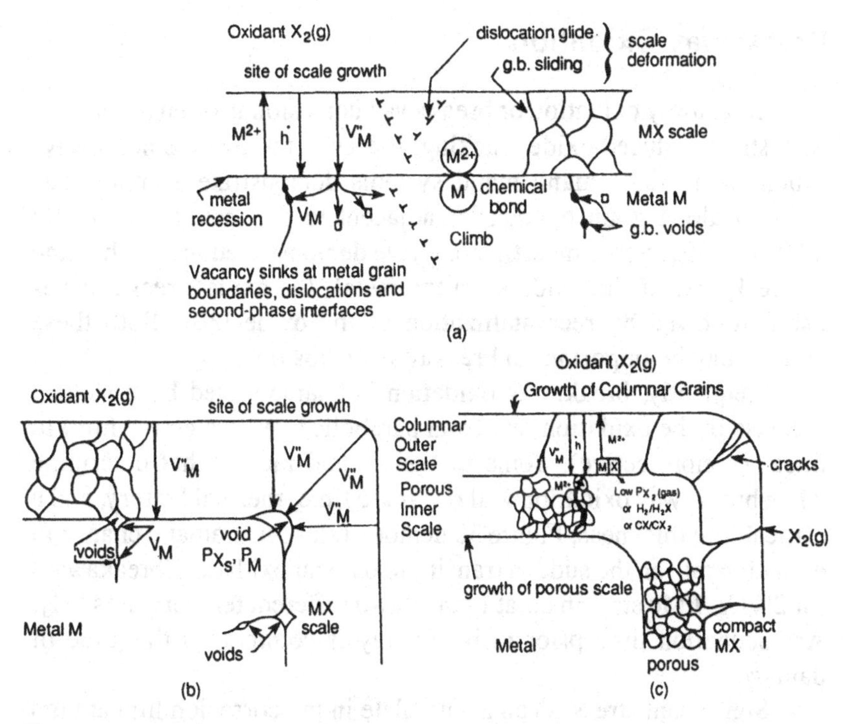

Figure 28. Problems occurring at the metal/film interface during the growth of cation-diffusing scales.

transfer component of corrosion and a diffusion component of corrosion. The two-component system is shown in Fig. 29 with the equivalent electric circuit [28]. Metal dissolution occurs at the metal/film interface, while oxygen reduction occurs at the metal/solution interface in the pore.

In the equivalent circuit, C is the capacitance of the electrode, Z_{me} is the charge transfer impedance, Z_n is the transport impedance (described by Nernstian diffusion of oxygen), $\sigma\bar{Z}_\sigma$ defines variations in linear transport ahead of the oxide layer, and $\sigma/(1+\sigma)\bar{Z}_{por}$ characterizes transport within the pores. While this model has been successful in some cases, it is only applicable when oxidation proceeds concurrently with oxygen reduction although at different sites, and the anodic reaction is principally charge transfer–controlled and the oxygen reduction is primarily diffusion-controlled.

Breakaway Oxidation

Breakaway oxidation or breakaway corrosion is often associated with stress-induced oxide cracking; however, the two are not always synonymous since many metal systems demonstrate a continuous porosity along grain boundaries adjacent to the metal/film (oxide) interface [29]. Some investigators have demonstrated that at the time of breakaway of the oxide from the metal, the oxide stress relieves itself, followed by recrystallization of the oxide [30]. Both these factors may be important in breakaway corrosion.

Frequently, breakaway oxidation is characterized by a sudden increase in the oxidation rate from parabolic (i.e., protective film) to linear (nonprotective). Some of the critical factors that determine when breakaway oxidation will occur are time, thermal history, and a critical film thickness. Figure 30 demonstrates the dramatic change in oxidation rate on the sudden transition to linear oxidation (breakaway) for 2¼ Cr–1 Mo steel in air at 1 atm at two different temperatures [31]. An incubation time prior to breakaway is common to this type of damage.

Significant stresses can accumulate in the corrosion film at high

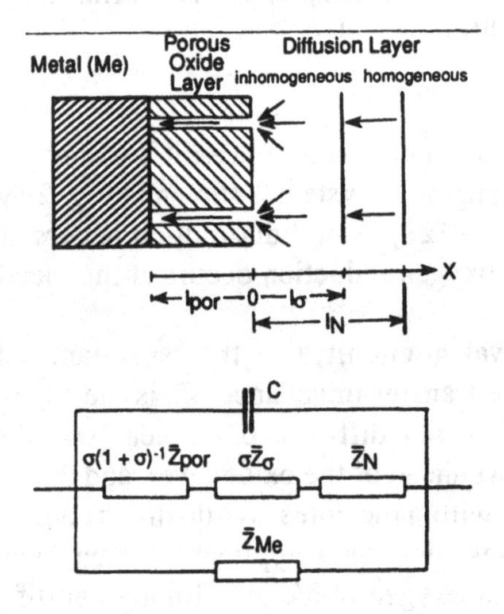

Figure 29. Inhomogeneous surface model of a porous film and the equivalent electric circuit.

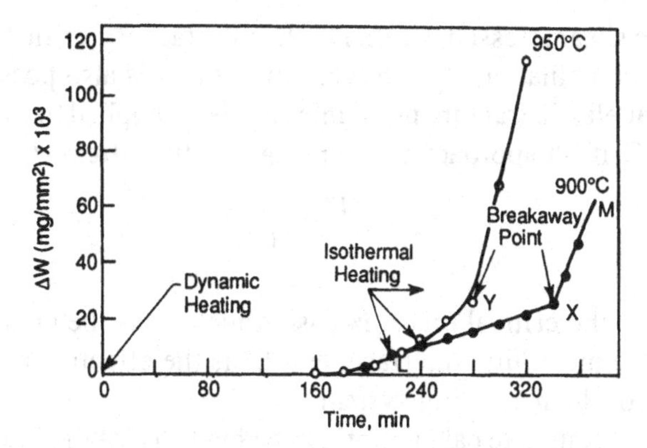

Figure 30. Thermogravimetric curves for the oxidation of 2¼ Cr–1 Mo steel in 1 atm air at 900°C and 950°C.

temperatures as a result of the different creep strengths of the metal substrate and the oxide. Moreover, if the metal/film system is cooled to lower temperatures, the difference in coefficients of thermal expansion between the film and the metal can result in large strains at the interface. These strains are in addition to the grown-in strain or stress in the film. While creep is not a consideration for films at low temperature, stress and film spalling are important even for corrosion generated in aqueous solutions.

Film Stresses and Decohesion

Oxides are usually less dense and more voluminous than the metals from which they form. For adherent films interfacial strains will exist at the metal/film interface that will produce stresses within the film itself. Furthermore, duplex or multiphase films will contain interfacial strains between each phase and likewise associated stresses. Of course, once decohesion or spalling of the film begins, rapid corrosion often follows due to easy access of the environment to the substrate.

Decohesion of the film from the metal substrate occurs when a critical stress is exceeded at the interface. If the stresses are of a tensile type perpendicular to the interface, the critical stress is σ_i, and if there are shear stressed parallel to the interface the critical stress is τ_i [32]. A combination of both types of stresses is also possible.

Linear elastic fracture mechanics has been applied to this system using the Griffith approach for a critical fracture stress σ_i:

$$\sigma_i = \left(\frac{E^* G_c}{\pi c} \right)^{1/2} \tag{16}$$

where G_c is the critical crack extension force, c is the critical crack length at the metal/film interface, and E^* is the effective modulus of elasticity for the metal/film system.

A more complete calculation can be made by taking into account the presence of voids along the separation plane. This factor can be included by replacing G_c with $(1 - V)\, G_c$, where V is the area fraction of voids [32]. Hence, substituting the stress intensity factor $K_I = \sigma(\pi c)^{1/2}$ into the equation (16) gives

$$K_{IC} = [E^*(1 - V)G_c]^{1/2} \tag{17}$$

The K_{II} stress intensity may likewise be determined for shear.

While this approach is useful for approximations and is applicable for truly elastic behavior, which may be the case for low-temperature films, high-temperature films are often quite ductile and sufficiently plastic that a linear elastic fracture mechanics approach may not be suitable. Moreover, at sufficiently high temperatures creep may become the predominant failure mode.

Mechanical failure of films does not always involve tensile or shear stresses. Compressive stresses have been shown to lead to the formation of oxide ridges (Fig. 31) [33]. Eventually buckling and fracture of the film occur, opening the substrate to the environment.

For multiphase films, the stress states become more complex. For example, Fig. 32 illustrates the relative magnitude and sign of the stress in each phase formed on 2¼ Cr – 1 Mo steel after cooling to room temperature [32]. In the first case (Fig. 32a), after short exposure to air at 600°C and then cooling down, decohesion occurs at the metal/film interface. After longer times at temperature (Fig. 32b), cracking occurs in the magnetite (Fe_3O_4) phase, and separation of the hematite (Fe_2O_3)/magnetite interface takes place.

The types and concentration of defects in films significantly

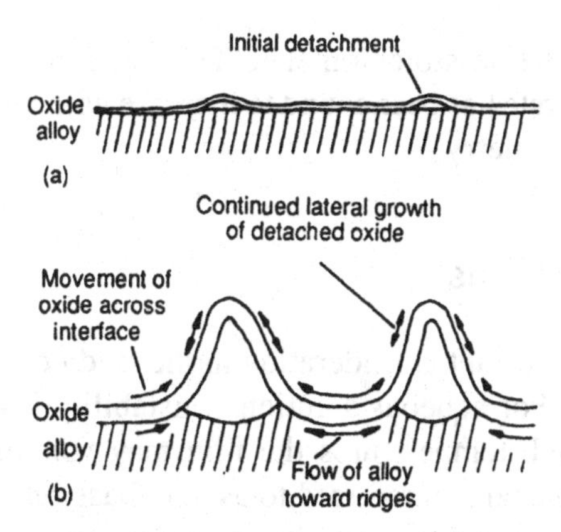

Figure 31. Illustration of lateral/film growth producing oxide ridges after (a) a few minutes and (b) several hours.

effects the mechanical behavior of the film [34]. For instance, the critical resolved shear stress of TiO_2 is greatly enhanced when it has been annealed to increase the concentration of anion vacancies. This increase in critical resolved shear stress will reduce the plasticity of the film, leading to premature fracturing of the film. This has been confirmed for nonstoichiometric TiO_2, which was found to be brittle

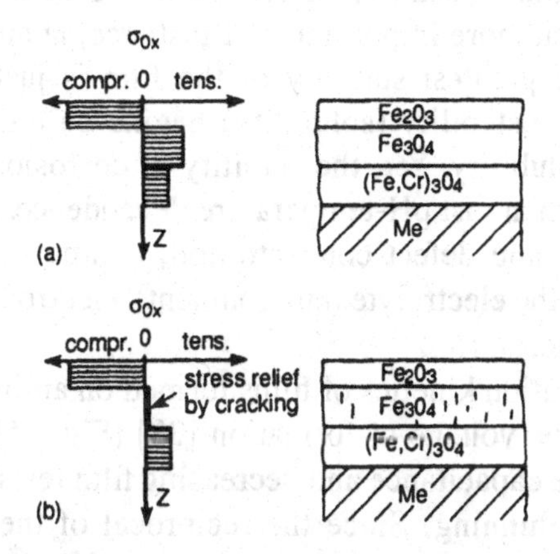

Figure 32. Schematic illustration of stress development in films on 2¼ Cr–1 Mo steel during cooling. (a) No cracking in magnetite during cooling; (b) microcracking in magnetite during cooling.

below 1000°C, while stoichiometric TiO_2 was quite ductile [35]. Likewise, creep rate has been found to increase with increasing deviation from stoichiometry.

Instability of Films

Another important consideration in the study of corrosion films is environment. For aqueous solutions, instability is determined by pH, while at high temperatures the film may volatilize or become unstable by a mechanism referred to as acid/base refluxing.

Fontana illustrated the stability of oxides formed on pure metals as a function of pH at 25°C based on Pourbaix diagram methods (Fig. 33) [36]. It is apparent that some oxides are stable over a large pH range (e.g., Co), while others have limited stability (e.g., Zn). At pH values on either side of the stability line, the oxide becomes soluble in the environment. Thus, there are conditions where the corrosion film formed in aqueous solutions may reach a dynamic equilibrium between continued film growth and dissolution while maintaining a quasi-steady-state film thickness.

Increasing solubility of the iron carbonate film formed on steel in CO_2 as a function of temperature is illustrated in Fig. 34 [37]. Although temperature has a large effect on the solubility of the $FeCO_3$ film, pH is much more important. For instance, at ambient temperature (25°C) the greatest stability of the film is just above pH 10; however, rapid instability (solubility) begins on either side of this minimum in solubility. Yet, the stability of corrosion films is much more complex than just pH-temperature dependence. Factors such as type of defects and defect concentration, composition of the film, composition of the electrolyte (environment), electrode potential, and current are important.

The dissolution kinetics of films formed on antimony are highly dependent on the voltage of formation [38] (Fig. 35). A decreasing recipricol of the capacitance and decreasing film resistance are indicative of oxide thinning. Since the reciprocal of the capacitance is proportional to the oxide thickness, the highest formation voltages produced the thickest films and thus took longer to dissolve. Color interferometry showed that oxide dissolution in sulfuric acid occurred

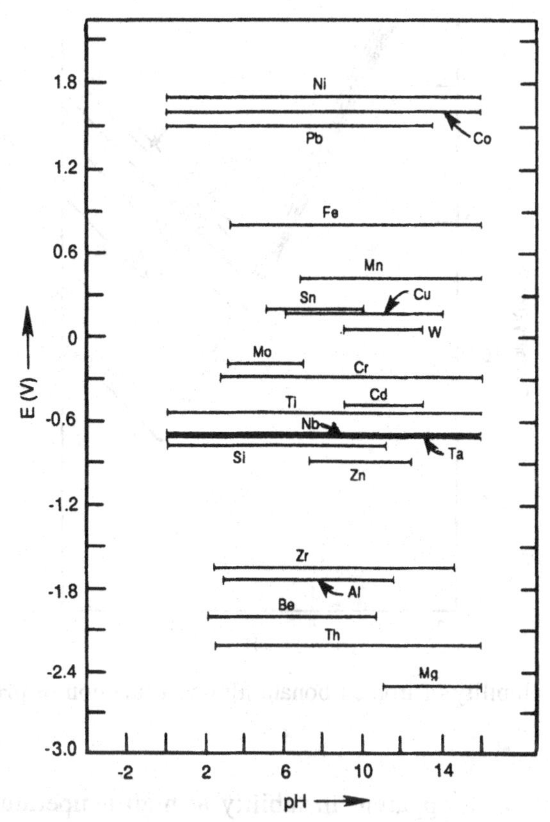

Figure 33. Maximum range of film stability on pure metals as a function of pH at 25°C according to Pourbaix diagrams, assuming 10^{-4} *M* soluble ions.

uniformly in reverse to oxide growth, indicating that film dissolution occurred essentially the same as film formation. The inflection in each curve may be attributed to the change in dissolution kinetics between two different composition oxides, probably Sb_2O_5 at the metal/film interface and Sb_2O_3 at the film/solution interface.

The high formation rate of antimony oxide in phosphoric acid and, thus, the correspondingly high dissolution rate compared to the film formed in sulfuric acid are expected to be the result of a greater concentration of defects in the former compared to the latter. Therefore, a higher defect population would enhance film dissolution [38] (Fig. 36).

Instability of the corrosion film is also a problem at high tem-

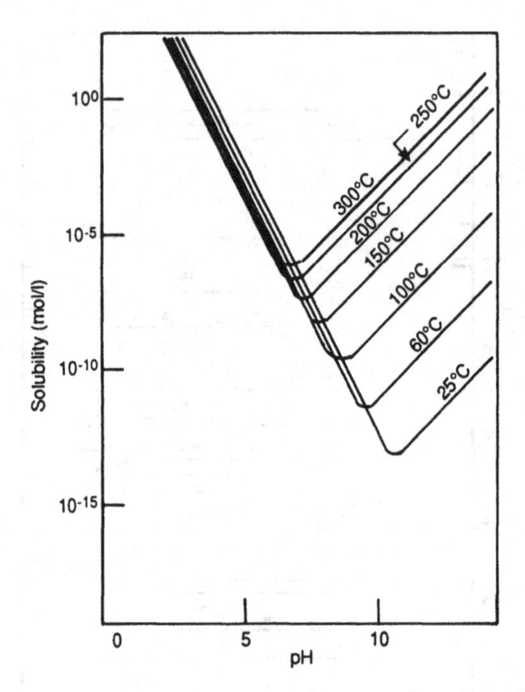

Figure 34. Solubility of iron carbonate film as a function of pH and temperature.

perature. The most apparent instability at high temperature is solubility of the film by phase change to a liquid. Once the temperature has exceeded the melting point of the oxide phase, instability occurs. However, two of the most important manifestations of film instability at high temperature, below the melting point, are oxide volatility and so-called hot corrosion. The former is observed for many materials, but Si-, Mo-, and Cr-based materials, where volatile species may form during oxidation, have been studied more extensively. Figure 37 shows several deviations in the parabolic rate law constants for Cr oxidation as a function of temperature [39]. The deviation at point C corresponds to volatilization of Cr metal to a vapor which could migrate through the oxide film at a faster rate than the metal ion. This volatilization of Cr metal to a vapor would occur at the solid metal/ $Cr_2O_3(s)$ interface according to thermochemical considerations. The additional change in slope at B is probably the volatilization of Cr_2O_3 to a gaseous phase at the $Cr_2O_3(s)$ film/oxygen environment interface.

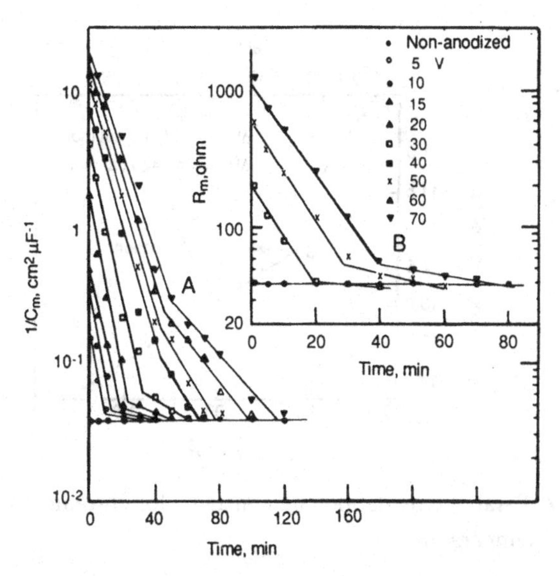

Figure 35. Decrease in reciprocal capacitance ($1/C_m$) and resistance (R_m) with time during dissolution of films formed on antimony.

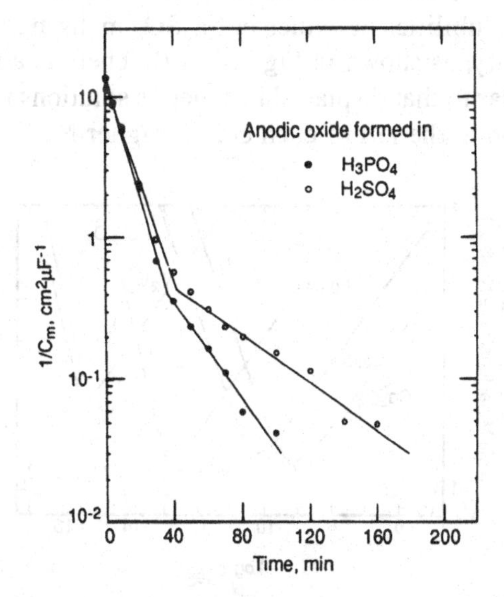

Figure 36. Variation of $1/C_m$ with time during dissolution of films formed on antimony in phosphoric and sulfuric acid solutions.

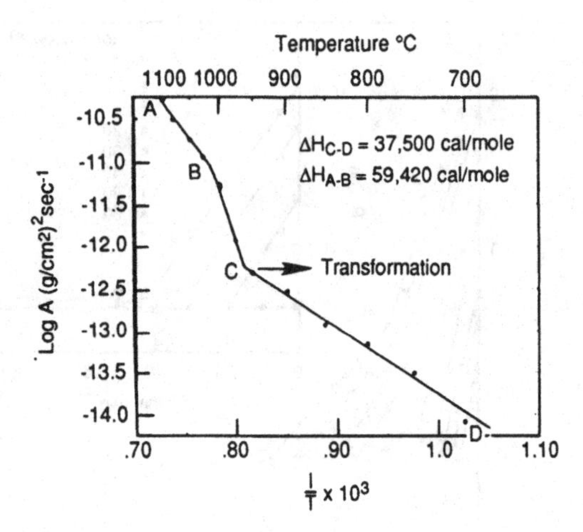

Figure 37. Parabolic oxidation rate constants for chromium in oxygen as a function of temperature.

The other film instability, often referred to as *hot corrosion*, is an increase in oxidation rate of a material due to the formation of a thin fused salt film. The salt most commonly involved in hot corrosion is Na_2SO_4. The solubilities of oxides in Na_2SO_4 melts are a function of the Na_2O activity, as shown in Fig. 38 [40]. There is a minimum in solubility similar to that displayed in aqueous solutions in Fig. 34. In fact, dissolution to the left of each curve (higher Na_2O activities) is

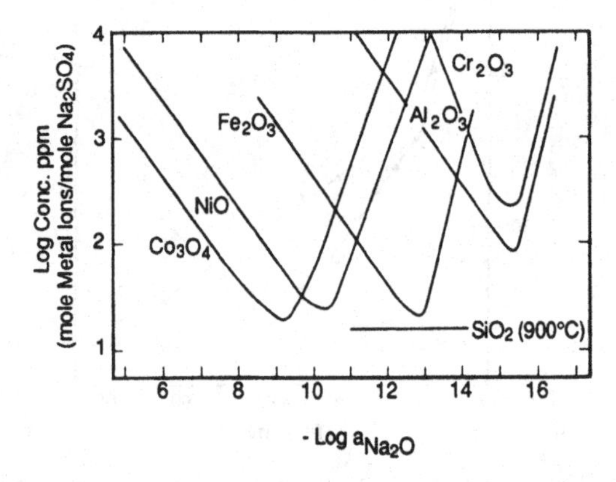

Figure 38. Measured oxide film solubilities in fused sodium sulfate at 1200 K and 1 atm oxygen.

referred to as *basic* dissolution, and that to the right of the minimum as *acidic* dissolution. Therefore, hot corrosion may involve the fluxing of corrosion films as either basic or acidic solutes in the Na_2SO_4 salt. Once the metal ions from the film are solubilized in the salt, they subsequently precipitate at a distance away from the metal/film interface as an oxide, leaving the metal unprotected and open to accelerated attack from the oxidizing environment.

References

1. M.G. Fontana, *Corrosion Engineering*, 3rd ed.; McGraw-Hill, New York, 1986.
2. J. Oudar, *Int. Met. Rev.* **1978,** Review 228, 57.
3. H.J. Leamy, G.H. Gilmer, and K.A. Jackson, *Surface Physics of Materials*, vol. 1, Ed. J. Blakely, Academic Press, New York, 1975.
4. J. Benard, *Oxidation of Metals and Alloys*, p. 6, ASM International, 1970.
5. S.A. Bradford, Corrosion, vol. 13, p. 61, *Metals Handbook*, ASM International, Metals Park, OH, 1988.
6. J.A.S. Green, *Corrosion* **1974,** *30,* 175.
7. Y.M. Wang and D. Radovic, Corrosion Research Symposium, Corrosion/88 NACE, St. Louis, MO, 1988.
8. R.A. Swalin, *Thermodynamics of Solids*, Wiley, New York, 1962.
9. F. Delnick and N. Hackerman, *J. Electrochem. Soc.* **1979,** *126,* 732.
10. K. Azumi, T. Ohtsuka, and N. Sato, *J. Electrochem. Soc.* **1987,** *134,* 1352.
11. K. Juttner, W.J. Lorenz, and F. Mansfeld, Corrosion/89, Paper No. 135, NACE, New Orleans, LA, 1989.
12. K. Azumi, T. Hotsuka, and N. Sato, *J. Electrochem. Soc.* **1989,** *134,* 1352.
13. K. Hauffe, *Oxidation of Metals*, p. 19, Plenum, New York, 1965.
14. J. O'M Bockris and A.K.N. Reddy, *Modern Electrochemistry*, vol. 2, Plenum, New York, 1977.
15. S.R. Morrison, *Electrochemistry at Semiconductor and Oxidized Metal Electrodes*, Plenum, New York, 1980.
16. A.G. Gad Allah and A.A. Mazhar, *Corrosion* **1989,** *45,* 381.
17. G.A. Parks, *Chem. Rev.* **1965,** *65,* 177.
18. T.D. Burleigh and R.M. Latanision, *J. Electrochem. Soc.* **1987,** *134,* 135.
19. W. Hirchwald, *Current Topics in Materials Science*, vol. 6, p. 115, North-Holland, Amsterdam, 1980.
20. J.M. Bastidas and J.D. Scantlebury, *Corrosion Sci.* **1986,** *26,* 341.
21. N. Sato, *Corrosion* **1989,** *45,* 354.
22. J. Paidassi, *Trans. Am. Inst. Min. Eng.* **1953,** *197,* 1570.
23. B.D. Cahan and C.T. Chen, *J. Electrochem. Soc.* **1982,** *129,* 474.
24. B.D. Cahan and C.T. Chen, *J. Electrochem. Soc.* **1982,** *129,* 921.
25. N.N. Khoi, W.W. Smeltzer, and J.D. Embury, *J. Electrochem. Soc.* **1975,** *122,* 1495.
26. A. Atkinson, R.I. Taylor; and A.E. Hughes, *High Temperature Corrosion*, Ed. R.A. Rapp, NACE, Houston, TX, 1983.
27. R.A. Rapp, *Met. Trans. A.* **1984,** *15*A, 765.
28. F. Mansfeld, *Corrosion* **1988,** *44,* 856.

29. B.J. Cox, *J. Nuc. Mat.* **1971,** *41,* 96.
30. D.H. Bradhurst and P.M. Hever, *J. Nuc. Mat.* **1971,** *41,* 101.
31. A.S. Khanna, B.B. Jha, and B. Raj, *Oxid. Met.* **1985,** *23,* 159.
32. M. Schutze, *Mat. Sci. Tech.* **1988,** *4,* 407.
33. F.A. Golightly, F.H. Stott, and G.C. Wood, *Oxid. Met.* **1976,** *10, 163.*
34. *D.L. Douglass, Oxidation Metal Alloys,* p. 137, Ed. D.L. Douglass, ASM, Metals Park, OH, 1971.
35. G.E. Hollox and R.E. Smallman, *Proc. Brit. Ceram. Soc* **1965,** *6,* 317.
36. M.G. Fontana, *Corrosion* **1971,** *27,* 129.
37. A. Ikeda, Sumitomo Metals, private communication.
38. M.S. El-Basioung, M.M. Hefuy, and A.S. Mogoda, *Corrosion* **1985,** *41,* p. 611.
39. E.A. Gulbransen and S.A. Jansson, *Oxidation Metal Alloys,* p. 63, Ed. D.L. Douglass, ASM, Metals Park, OH, 1971.
40. R.A. Rapp, *Corrosion* **1986,** *42,* 568.

4

Corrosion Inhibition and the Role of Films

Introduction

Corrosion inhibitors are chemical agents that effectively reduce corrosion of a metal surface by reacting with the metal surface directly, called *interface inhibition*, by reacting with or modifying a preexisting corrosion product film, called *interphase inhibition*, or by scavenging constituents from the solution before they react with the metal. The latter group of inhibitors will not be dealt with in this chapter. Moreover, only the mechanisms of chemical inhibition in conjunction with corrosion films will be discussed. Neither the chemistry of the inhibitors themselves, nor such molecular factors as geometrical dimensions and molecular structure, cross-sectional area, π and σ bonding, ionization potential and electron affinity, spatial relationships between different functional groups, dipole moments, etc., are discussed. All of these factors and others are important to the mechanisms and theories of inhibition by inhibitors; however, they represent another aspect of corrosion inhibition not germane to this text.

Inhibition Models

Interface inhibition is generally defined as adsorption of the inhibitor directly onto the metal surface. This type of inhibition is most likely to occur in acidic solutions where oxide films are not stable, and thus inhibitors have direct access to the surface through the double layer. Likewise, nonaqueous solutions may provide environments where a stable oxide is not present and interface inhibition is the mechanism. The adsorption of these inhibitors is often on the order of a monolayer or less in thickness, and therefore they are represented as a two-dimensional film.

Interface inhibition probably occurs by one of the following means: geometric blocking, deactivating coverage, or reactive coverage [1]. *Geometric blocking* is essentially the creation of a physical barrier by a high degree of coverage with the inhibitor. This is the more classical model of a corrosion inhibitor, where the inhibitor produces an inert obstacle to exposure of the metal surface to its environment. Therefore, the reaction rates can be expected fo fall in proportion to the extent of surface coverage. This behavior will result in a downward shift of the anodic part of the polarization curve with no corresponding change in Tafel slope. Figure 1 [2] illustrates this behavior for steel in 0.5 *M* sulfuric acid with various concentrations of the inhibitor TPBP+. The dissolution kinetics of iron continue unchanged at those sites not covered by the inhibitor. As the surface coverage increases with increasing concentration of inhibitor, the

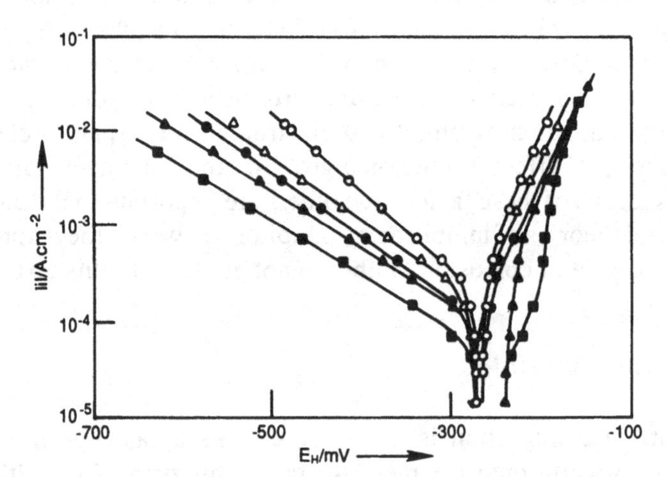

Figure 1. Steady-state galvanostatic current density versus potential at 25°C. Iron in deaerated 0.5 *M* H_2SO_4 + *mM* TPBP+, where (○) $x = 0$, (Δ) $x = 0.1$, (●) $x = 1.0$, (▲) $x = 5$, (■) $x = 10$.

anodic curve moves downward to lower corrosion currents yet maintains a constant slope. Figure 2 shows the same geometric blocking behavior for 1060 aluminum in a variety of benzoic acid inhibitors [3]. Here the cathodic curve slope remains essentially constant.

Deactivating coverage requires adsorption only at active sites on the surface; blocking the necessary reactions. Thus, inhibition of either the reduction or oxidation reactions or both may occur. This coverage is, of course, lower than for the geometric blocking inhibitors. The deactivating coverage inhibitors are generally inert. Figure 3 [2] demonstrates the deactivating of active sites on iron in 0.5 *M* $HClO_4$ by the inhibitive action of $Pb(ClO_4)_2$. The shift from a measurable Tafel slope for the anodic reaction in the absence of inhibitor to essentially no Tafel slope with 10^{-3} *M* inhibitor indicates the active sites for iron dissolution have been successfully blocked or deactivated.

The third type of interface inhibitor entails a *reactive coverage*, where the inhibitor is not inert but participates in oxidation or reduction reactions to reduce the overall corrosion rate. Moreover, the inhibitor itself may be reduced or oxidized to induce secondary inhibition. An example of this type of interface inhibition has been found

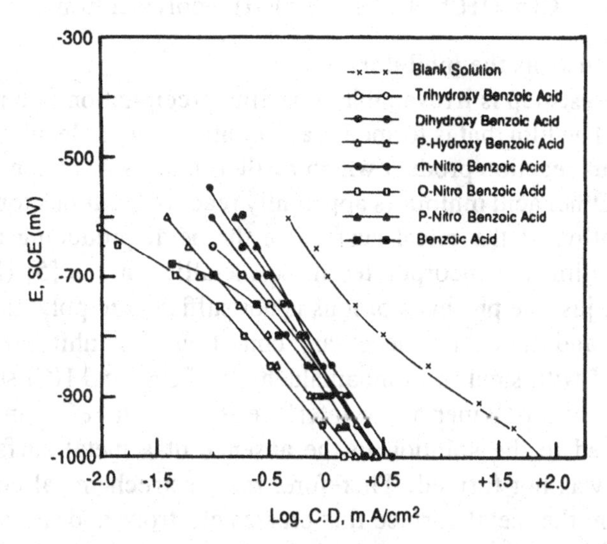

Figure 2. Cathodic polarization curves for 1060 aluminum at 35°C in 0.5% inhibitor of various composition.

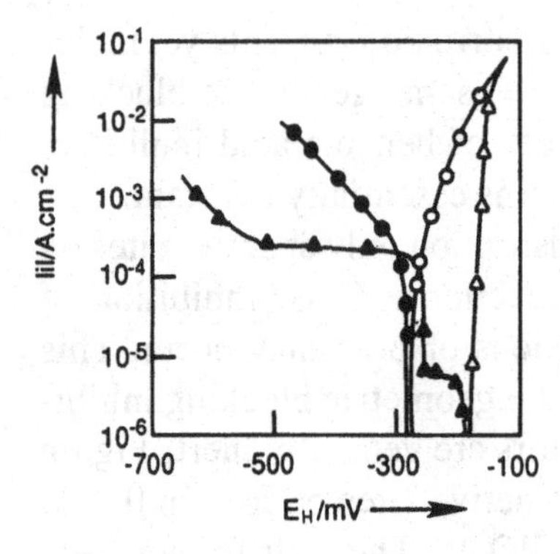

Figure 3. Steady-state current density versus potential for iron in 0.5 *M* $HClO_4 + yM\ Pb(ClO_4)_2$ at pH 0.3 and 25°C. $y = 0$ (○) anodic, (●) cathodic $y = 10^{-3}$ (△) anodic, (▲) cathodic.

for octynol, 3-phenyl-2-propyn-1-ol and 1-phenyl-2-propyn-1-ol used to inhibit steel exposed to strong mineral acids. These inhibitors are protonated, then adsorbed to the metal surface, and finally react to form a polymer film that inhibits transport of ions to or away from the metal surface [4]. This behavior has been reduced to reaction form for steel in HCl as follows:

$$I + H^+ = IH^+$$
$$Cl^- + IH^+ = (Cl^-)(IH^+)$$
$$(Cl^-)(IH^+) + nI + (Cl^-)(H) \rightarrow \text{polymer film}$$

Here I represents the inhibitor.

The last step is irreversible reflecting precipitation of a reaction product. The film that is formed is apparently a low-molecular-weight amorphous reaction product which further inhibits diffusion of ionic species. Other acid inhibitors apparently react with cations formed by the oxidation of the metal surface in the acid, producing a stable inhibitor film that incorporates these metallic cations [5] (Fig. 4). However, just the presence of ions is not sufficient to polymerize the inhibitor and create the necessary conditions for inhibition. Tests conducted with steel in *t*-cinnamaldehyde (TCA) and HCl show the formation of a polymer at the interface [6]. When $FeCl_2$ and $FeCl_3$ were added to the solution in the absence of a metal surface, the polymer was not formed. Therefore, the electrochemical corrosion process on the metal surface that creates electrons and the resulting potential differences are required to induce the polymerization process necessary for inhibition. Thus the properties of the polymeric

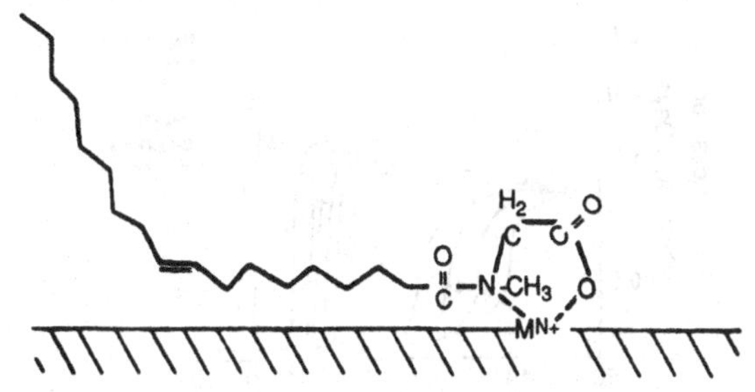

Figure 4. Surface interaction of *N*-oleoylsarcosine with metal ion such as iron.

film become controlling, much the same as for oxide films. However, little is known about the ionic and charge transport through these organic films.

Interphase inhibitors are considered to form a three-dimensional phase rather than a two-dimensional structure like the interface inhibitors. The three-dimensional interphase may be a solid, such as a reaction product, adsorbed compounds, or insoluble compounds, or a liquid film that has properties different from the electrolyte.

One of the earliest known classes of interphase inhibitors that possess the ability to promote stable oxide formation is the *passivators.* These inhibitors shift the corrosion potential into the passive range so that a stable oxide is formed, thereby reducing corrosion. Therefore, only metals that demonstrate an active-passive transition are inhibited by passivators. Passivators, also known as *oxidizing inhibitors,* are most effective in neutral or slightly acidic electrolytes. Some of these inhibitors also require the presence of oxygen to be effective; others do not. For example, molybdates and tungstates fall in the category of requiring oxygen, while nitrites and chromates induce passivation with or without oxygen. The passivating inhibition of some of these anions is evident in Fig. 5 [7]. Under neutral conditions with aeration, molybdates are the most effective inhibitors for iron; however, changes in pH, temperature, solution chemistry, and alloy content can significantly alter these relationships.

For molybdates inhibition appears to be the result of some form of molybdenum incorporated in the growing oxide film [8]. While the evidence for molybdenum incorporation is weak in iron-base alloys, it has been confirmed for aluminum, zinc, and tin. These films are

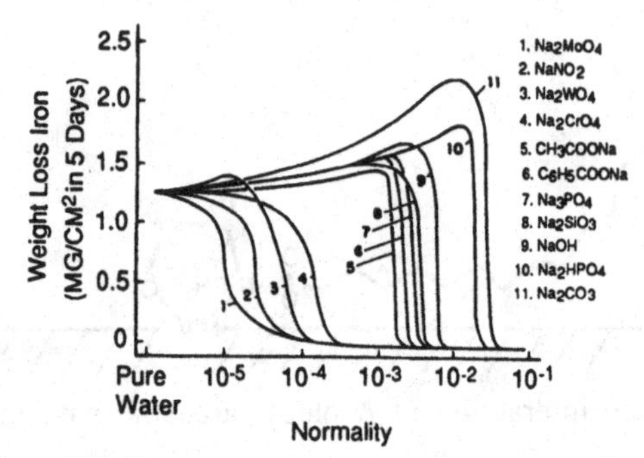

Figure 5. Effect of inhibitor composition and concentration on corrosion of iron in aerated aqueous solution.

apparently composed of insoluble ferric molybdates for ferrous alloys and aluminum molybdates for aluminium alloys. The beneficial role of MoO_4^{2-} on passive films of austenitic stainless steel may be the results of reversing the anion-selective nature of the passive film to cation-selective, thereby reducing the incorporation and transport of chlorides.

Chromates are definitely incorporated in the oxide film, as shown in Fig. 6, from x-ray photoelectron spectroscopy (XPS) analysis [9]. However, XPS additionally indicates that chrome within the film is actually reduced from the 6^+ valency to the 3^+ valency during adsorption of chromates. The low M/Fe ratio for CrO_4^{2-} (Table 1) confirms this behavior, which is contrary to the role of other oxidizing inhibitors in Table 1 [10]. The M/Fe ratios greater than 1 for all the other passivators indicate that corrosion protection is not achieved by

Table 1[a]

Na_xMO_4 inhibitor	MoO_4^{2-}	VO_4^{3-}	VO_3^-	WO_4^{2-}	CrO_4^{2-}
Protective concentration, g/L	5	1	10	1	0.1
Energy E_B $Fe2p_{3/2}$[b]	711.4	711.6	711.6	711.2	711.6
Binding energy E_B M	235.5	517.4	517.4	35.8	576.8
M/Fe ratio	11	6.3	2.3	2	0.5
O/Fe ratio	9	16	11	12	30

[a]Integral peak intensities used including coefficient effect σ.

[b]$Fe2p_{3/2}$ binding energy is 211.2 eV in Fe_2O_3, 710.6 eV in Fe_3O_4; $Cr2p_{3/2}$ is 576.9 eV in Cr_2O_3, 579.3 eV in $Fe_2(CrO_4)_3$.

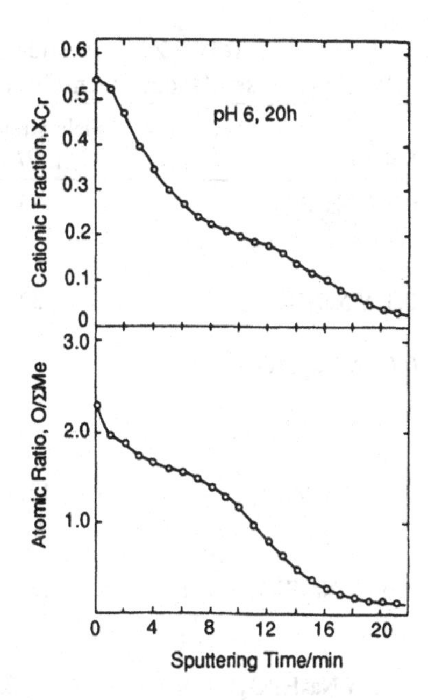

Figure 6. Composition versus depth profile for passive film on iron formed in aerated 0.01 M Na_2CrO_4 solution at pH6.

large concentrations of inhibitor. In fact, for some inhibitors, such as pertechnate, as little as 4% surface coverage on steel has been found sufficient to completely arrest corrosion. This contradicts the old concept of inhibitors as barriers. If inhibitors did protect by some barrier property, they would have to cover a much larger surface area, approaching or exceeding unity. For some of the interface inhibitors in acid solutions where polymerization occurs, the barrier effect may occur.

Besides the reduction of chrome in iron oxide films there are data that show that the increase of the chromium/iron ratio in the film eventually alters the conductivity from *n*-type to *p*-type. The *p*-type conductivity is more desirable for pitting resistance.

The protectiveness of films formed in the presence of passivators has been related to the electric field strength of the oxide formed during exposure to the inhibited solution. Table 2 compares the field strengths and growth rate of films formed on iron in aqueous solution exposed to air [11]. Those inhibitors that produced the highest field strength and the lowest growth rate were the most protective, while low field strength and long growth rates produced films of limited protection. The most protective films were found by Auger to be

Table 2. Properties of Film Formed on Iron While Exposed Under Open-Circuit Conditions to Inorganic Inhibitors

Solution	pH	Field strength, mV/Å	Growth rate *R*/decade time	Auger analysis
0.1 *N* Na_3PO_4	12	35	5	O, Fe, P Only on the surface
0.1 *N* $NaNO_2$	7	32	10	O, Fe, Small amount of N
0.01 *N* Na_2CrO_4	8–12	3–28	20–33	O, Fe, Cr near the substrate lack of Cr
0.1 *N* Na_2MoO_4	9.6	13.5	15	O, Fe, small amount of Mo
0.05 *N* Na_2WO_4	8.5	13.5	15	O, Fe, substantial amount of W
0.01 *N* Na_2HPO_4	9.1	0.02	Several thousand	O, Fe, P

composed entirely of iron oxide, while the less protective films contained ions from the solution and took longer to achieve protective potentials. Therefore, the longer film growth rates allow for more ferrous ion diffusion away from the metal/film interface (higher corrosion) than shorter film growth rates, which reduce ion diffusion from the interface.

The protective iron oxide is consistent with the frequently observed duplex film on steel of Fe_3O_4 adjacent to the metal surface and γ-Fe_2O_3 adjacent to the solution. These compositions are found in the presence of sodium nitrite, sodium phosphate, and sodium chromate. Similar compostions are observed with other organic and inorganic inhibitors, yet the degree of protectiveness varies significantly, which implies film composition is not always of primary importance but that film structure, electric field strength, and defect structure may be more important.

Quantum chemical calculations performed to determine energy conditions of passivators in the adsorbed state have confirmed XPS data that certain MeO_4^{2-} inhibitors act as electron acceptors. Moreover, these inhibitors do not function by direct metal ion/passivator interaction but by accepting electrons from the oxide itself. Additionally, it has been observed that the adsorption of some oxidizing inhibitors on oxide films reduces the potential drop across the film. This reduction in field strength would also limit further the driving force for cation diffusion and, thus, corrosion. However, this is contrary to the results in Table 2.

Photoelectric polarization has been used to analyze the defect response of oxide films to the application of passivators. The photoelectric response amplitude, V_{PEP}, is related to the ratio of oxide defects by the expression [8]

$$V_{PEP} = \frac{kT}{e} \ln \frac{X_c^{m+}}{X_a^{n-}}$$

where k is Boltzmann's constant, T is absolute temperature, e is the charge on an electron, X_C^{m+} is the concentration of cation vacancies with m^+ valency, and X_a^{n-} is the concentration of anion vacancies with n^- valency.

A positive value of V_{PEP} indicates an excess of cation vacancies, while a negative value of V_{PEP} indicates an anion vacancy excess. When V_{PEP} becomes zero, the oxide has reached stoichiometry with properties tending toward an insulator. The greater the value of V_{PEP}, the wider is the deviation from stoichiometry.

Figure 7 illustrates this affect for the application of sodium tungstate to steel in $NaSO_4$ solution [10]. When no inhibitor is added, the photoeffect amplitude is high and positive (curve 1), indicating a large

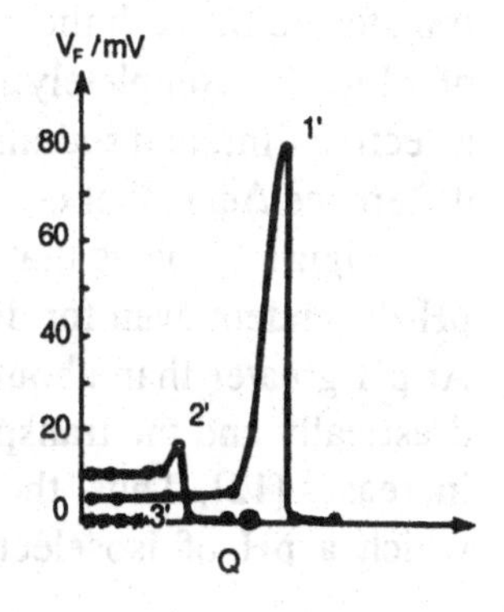

Figure 7. Photoelectric signal response (V_F) of steel as a function of charge transfer and concentration of Na_2WO_4 under anodic polarization: 1′—1 N $NaSO_4$ solution, $i_a = 300$ mA/cm²; 2′ + 3.3 g/L Na_2WO_4, $i_a = 133$ mA/cm²; 3′ + 16.5 g/L Na_2WO_4, $i_a = 7.7$ mA/cm².

deviation from stoichiometry and an excess of anion vacancies. The introduction of Na_2WO_4 decreases the photoresponse (curve 2) until at a concentration of 16.5 g/L stoichiometry of the oxide is achieved and cation vacancies and anion vacancies reach parity. These data are strong evidence for modification of both the electronic and ionic components of the film by an oxidizing inhibitor versus simply altering the composition, thickness, or protectiveness of the film. Furthermore, it demonstrates the importance of analyzing the properties of the film in conjunction with inhibitor studies rather than simply reporting the composition, which often does not change with inhibitor additions.

Another concept for the mechanism of inhibition by passivators, as well as many other inhibitors, is through alteration of the ion selectivity of corrosion films. It has been suggested that passivators adsorb at the film/solution interface, changing the ion selectivity of the film from anion-selective, which enhances migration of chloride ions, for example, to cation-selective, which reduces the corrosion rate. This film modification is supposedly accomplished by shifting the *potential of zero charge* (PZC) upon adsorption of the inhibitor. The potential difference of the film/solution boundary is apparently changed by adsorption of the inhibitor.

Figure 8 illustrates the change in ion selectivity of a ferric hydroxide membrane with the introduction of molybdate ions [9]. C_1 represents the sodium chloride concentration on one side of the membrane, which is held constant at 0.01 *M*. C_{11} is the bulk sodium chloride concentration on the other side of the membrane. In a neutral chloride solution the ferric hydroxide film displays anion-selectivity, as noted by the negative slope. Thus, chloride ions are preferentially transferred through the hydroxide membrane. However, the addition of MoO_4^{2-} completely reverses this behavior, producing a cation-selective film, and sodium is preferentially transferred. The potential difference $\Delta\phi$ is that developed across the membrane.

Figure 9 shows that the ion selectivity of this corrosion film is pH-dependent even for different anion species (i.e., OH^- and Cl^-). At pH greater than about 10.3, the transport of chlorides decreases drastically and the transport of hydroxide ions across the membrane increases [12]. Thus, there is a pH dependence of ion selectivity for which a pH of isoselectivity exists (identified as pH_{ps}) when the

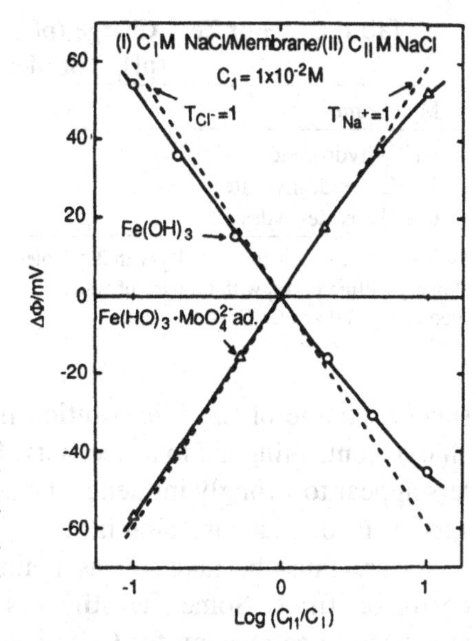

Figure 8. Membrane potential across hydrated ferric oxide membranes with and without adsorbed MoO_4^{2-} ions. The concentration of NaCl in compartment 1 is constant at 0.01 *M*.

membrane is nonselective. While this pH_{ps} is similar to the pH of zero charge (pH_{zc}), they are not identical. Table 3 presents the two different parameters for nickel, iron, and chromium membranes [13]. The

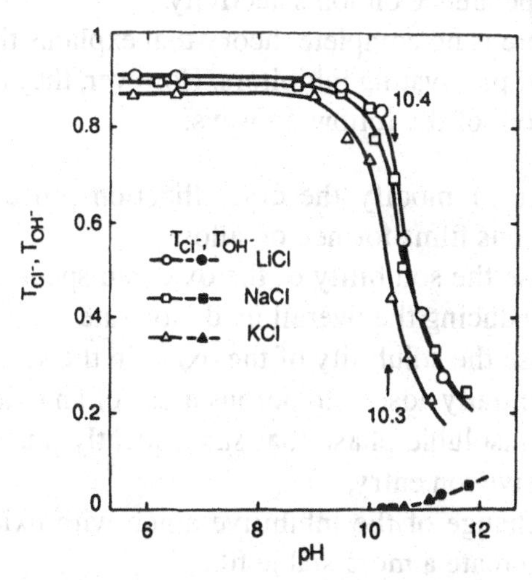

Figure 9. Variation of transport number in hydrated ferric oxide membrane as a function of pH.

Table 3. pH of Zero Charge (pH_{zc}) and Point of Isoselectivity (pH_{ps}) of Membranes

Membrane	pH_{zc}	pH_{ps}
Ni(II) hydroxide[a]	9.6–11.3	>13
Fe(III) oxide-hydrate[b]	8.6	10.3
Cr(III) oxide-hydrate[c]	6.5–7.4	~11

[a]Poorly crystallized beta $Ni(OH)_2$ with 3.5 moles of water.
[b]Noncrystalline Fe_2O_3 with 6 moles of water.
[c]Poorly crystallized Cr_2O_3 with water.

surface charge of the film/solution interface is not the same as the charge controlling the ionselectivity. However, both of these parameters appear to strongly influence the adsorption and transport of ionic species through a corrosion film.

Care must be taken in assigning ion selectivity properties to corrosion films. Some investigators have simply presented ESCA (Electron Spectrascopy for Chemical Analysis) data to demonstrate enhanced selectivity because of increasing concentration of an element in the film. While these films have ionic and electronic aspects, they are also subject to simple diffusion gradients that may enhance the concentration of elements in the film from the metal or the solution with no dependence on ion selectivity.

As yet, there is no complete theory that explains the mechanism of MeO_4^{2-}-type passivating inhibitors. However, they are believed to act in one or more of the following ways:

- ☐ Enhance or modify the crystallization process of initially amorphous films formed on alloys
- ☐ Decrease the solubility of the oxidized species (i.e., ferrous ions), reducing the overall oxidation rate
- ☐ Decrease the solubility of the oxide in the solution
- ☐ Preferentially adsorb to porous areas of an oxide film, creating an insoluble phase that subsequently plugs the pores to aggressive ion entry,
- ☐ Ion exchange of the inhibitive anion with oxide ions in the film to create a more stable film.

The pore-plugging model suggests that the inhibitive anion reacts with metallic cations in the vicinity of the pore, precipitating an insoluble product that stifles further corrosion in the pore. Pore plugging has been observed under conditions that are unfavorable for inhibition. Furthermore, the pore-plugging mechanism is most likely to occur when local pore pH of the solution is neutral or basic, promoting precipitation of solid reaction products.

It has been proposed that sodium phosphate inhibits by the pore-plugging mechanism to form ferric phosphate precipitates by the following reactions:

$$Fe^{2+} + H_2PO_4^- \rightarrow FeH_2PO_4^+$$
$$FeH_2PO_4^+ + 2H_2O \rightarrow FePO_4 \cdot 2H_2O + 2H^+ + e$$

Sufficient data are not available to confirm this mechanism. Not all metals inhibited by oxidizing inhibitors form insoluble precipitates; therefore, this model is not all inclusive. The other mechanisms for inhibition have been addressed in this chapter, but none are sufficiently developed to be considered a complete model.

In a similar fashion to inorganic passivating-type inhibitors, some organic inhibitors have been found to incorporate themselves within the corrosion film (interphase) to produce inhibition. Figure 10 is a schematic representation of the role of an organic inhibitor used to protect steel from corrosion by H_2S [14]. The corrosion product that often forms under these conditions is mackinawite, which is a non-stoichiometric iron sulfide (Fe_9S_8, more properly $Fe_{(1+x)}S$) [15]. The

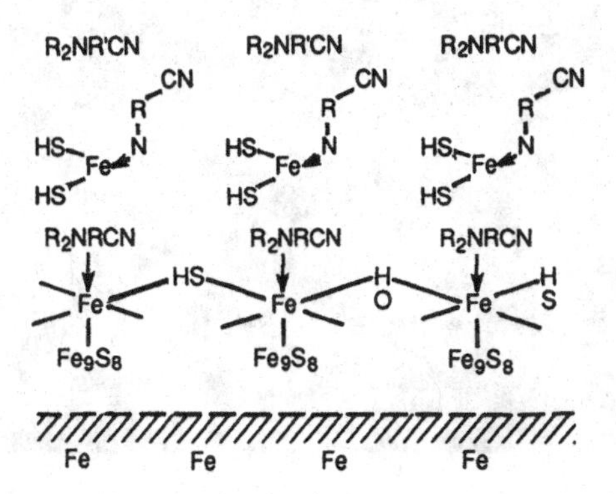

Figure 10. Protective iron sulfide film formed on steel in the presence of an organic inhibitor.

organic compound is observed to interact with iron ions and incorporate the corrosion film to provide inhibition.

Recent work has demonstrated that these organic inhibitors can significantly alter the structure of corrosion films. Figure 11 shows the structure of the surface film on uninhibited low-carbon steel exposed to an environment containing H_2S and CO_2 [16]. Under the same exposure, but in the presence of a 5000-ppm inhibitor, the morphology of the corrosion film has significantly changed (Fig. 12) [16]. This observation is contradictory to the classical concept of organic film-

Figure 11. Crystallinity of corrosion film on steel in 10% H_2S, 4% CO_2 at 220°C, 2000× magnification.

ing inhibitors acting as a hydrophobic barrier that completely covers the metal surface and acts as a corrosion barrier. Furthermore, strong evidence exists that indicates significantly less inhibitor than required for monolayer coverage is needed to produce complete inhibition. Therefore, the role of many of these inhibitors is surely to alter the structural, ionic, and electronic characteristics of the corrosion film in which they integrate.

As in the earlier-described case for octynol on steel in interface inhibition, organic inhibitors may react with the surface oxide to

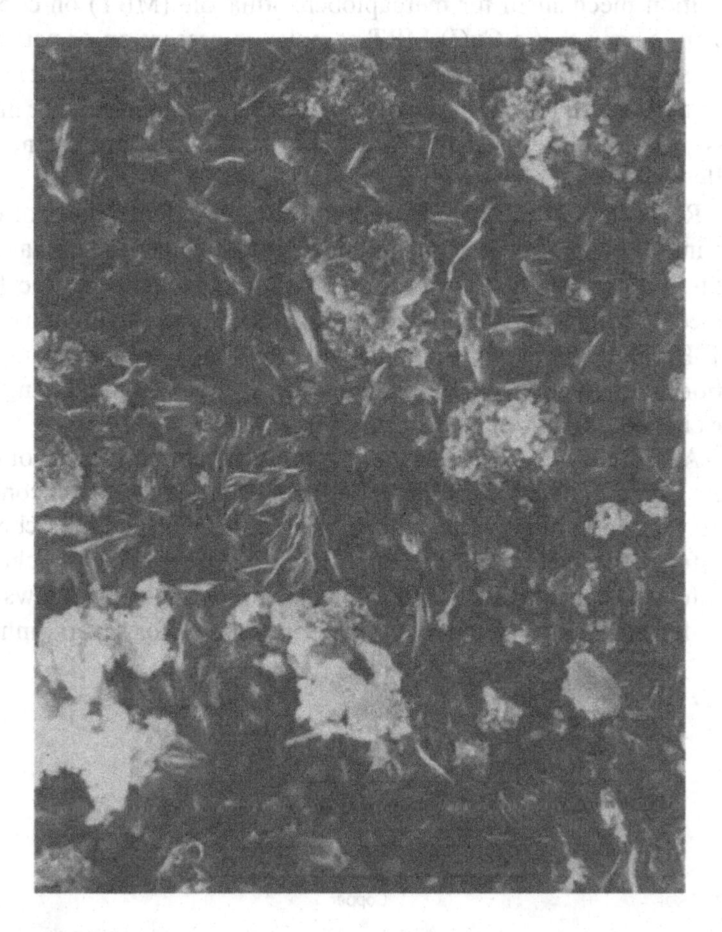

Figure 12. Change in crystallinity under same conditions as Fig. 11 with 5000 mg/L inhibitor added, 2000 ×.

produce a protective polymetric film. Benzotriazole (BTA) is suspected of just such behavior on copper, where cuprous oxide assists the formation of a protective polymeric triazole film of the type [Cu(I) BTA]. However, unlike the interface inhibition this represents three-dimensional interphase behavior. Other work using surface-enhanced Raman scattering has shown the presence of adsorbed BTA ions on the metal surface rather than a neutral polymeric compound. Moreover, evidence exists to support adsorption with cuprous and/or cupric ions and their oxides. Figure 13 illustrates one concept of the inhibition mechanism for mercaptobenzothiazole (MBT) on copper [17]. It is seen that a Cu(I) MBT complex is present on or partially incorporated into the cuprous oxide formed on Cu. This conflict in the exact nature of BTA inhibition of copper, whether interfacial or interphase, demonstrates that the two types of inhibition are not necessarily distinct but may overlap.

Regardless of the specific mechanism, the effectiveness of certain inhibitors can be established through AC-impedance measurements. Figure 14 illustrates the decreasing capacitance of the film formed on copper as a function of potential and inhibitor composition [18]. Here the mercaptobenzimidazole offers the best corrosion protection at all potentials of all the inhibitors evaluated by reducing the capacitance of the film.

Again, using AC impedance, the inhibition effectiveness of different organic inhibitors (filming amines) used for corrosion control in oil and gas production was determined and presented as charge transfer resistance across the inhibitor film: the greater the charge transfer resistance, the lower the corrosion rate. Figure 15 shows the time dependence of the charge transfer resistance for several inhibi-

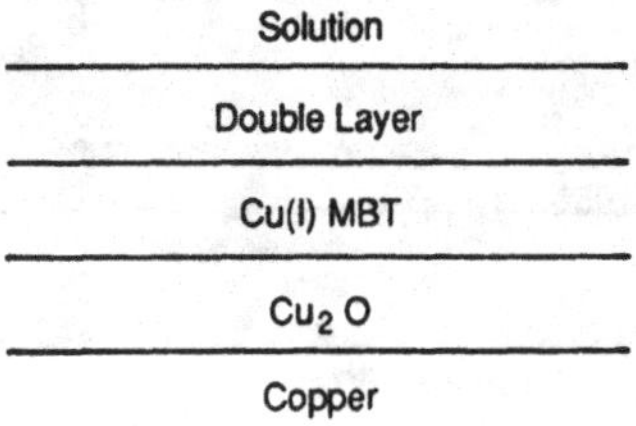

Solution

Double Layer

Cu(I) MBT

Cu_2O

Copper

Figure 13. Schematic of the phases formed during the inhibition of Cu in MBT.

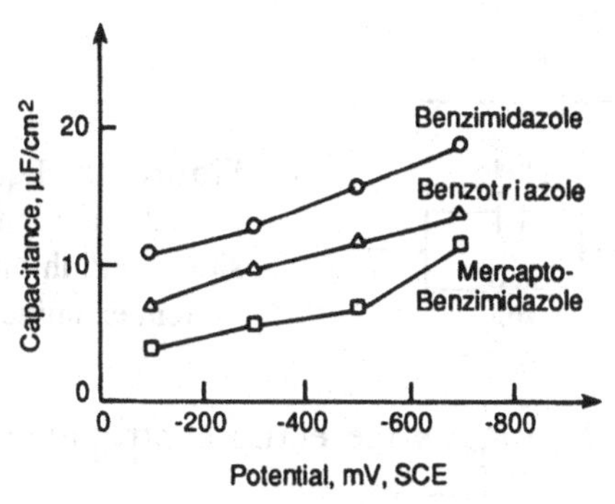

Figure 14. Capacitance versus potential for Cu in 0.1 *M* NaCl as a function of inhibitor composition.

tors compared to bare steel [19]. Inhibitors A and B demonstrated inhibition behavior several orders of magnitude greater than the bare steel and inhibitors C and E.

An equivalent circuit can be used to illustrate the metal/inhibitor system [20] (Fig. 16). The film capacitance, C_{pf} and the polarization resistance of the film, R_{pf}, represent the properties of the adsorbed inhibitor film, while R_o is the ohmic resistance of the cell and C_{dl} and R_p are the double-layer capacitance and polarization resistance, re-

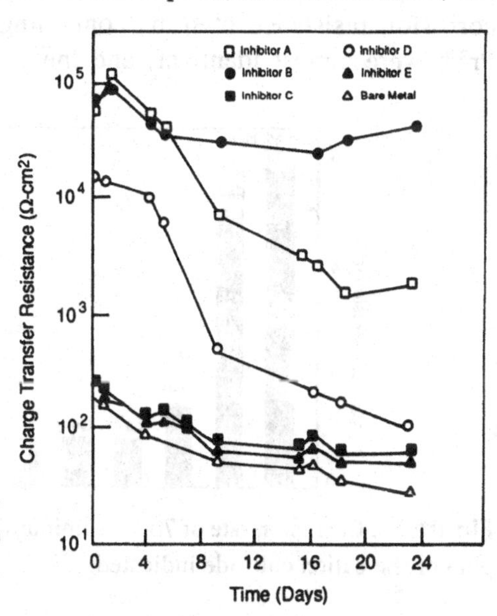

Figure 15. Charge transfer resistance as a function of time for steel in various inhibitors.

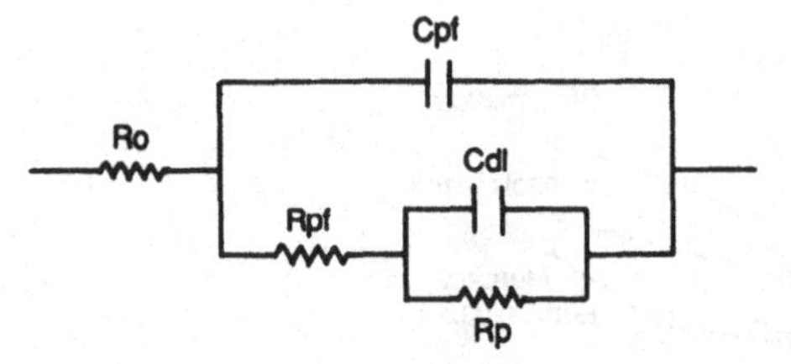

Figure 16. Equivalent electrical circuit for the steel/corrosion film/inhibitor/double-layer system examined in Fig. 15.

spectively. Thus, $R_p + R_{pf}$ is the entire contribution to the charge transfer resistance.

Modification of the corrosion film by either ion implantation of the metal surface or addition of specific cations to the solution can also inhibit corrosion. For the latter case, Fig. 17 is an example of the benefit of certain cations in solution on the corrosion inhibition of 7075 aluminum exposed to 0.1 *M* NaCl [21]. As interaction of the alumina film with the solution takes place, these ions become incorporated in the film. Auger analysis confirmed these cations were incorporated in the corrosion film.

Generally, the cations were present as metal oxide/hydroxide compounds, indicating they preferentially form at cathodic sites where OH^- are in abundance. This approach is not suitable for low-pH solutions where the film would be soluble in its environment. The corrosion resistance of films containing Nd^{3+}, Ni^{2+}, Ce^{3+}, and Pr^{3+} were almost identical, and the film compositions and thick-

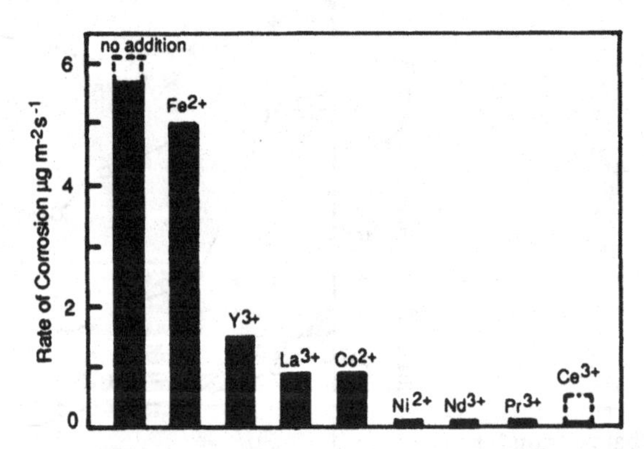

Figure 17. Corrosion rate of 7075 aluminum in 0.1 *M* NaCl containing 1000 ppm of the cation chloride indicated.

nesses were similar. Thus the mechanism by which these cations inhibit corrosion of 7075 aluminum in NaCl appears to be the same, probably through modification of the charge transfer characteristics of the film or the electric field strength.

Most of the discussion thus far has centered on inhibition in acid and neutral pH solutions. In solutions with basic pH, the mechanism of inhibition is still predominantly through interphase behavior, principally due to the abundance of hydroxyl ions. Increasing OH^- in solution near the metal surface, especially adjacent to cathodes, will reduce the solubility of cations and lead to the precipitation of a metal hydroxide deposit. Some metals that form stable oxides and hydroxides at high pH are cobalt, nickel, chromium, manganese, zinc, cadmium, cerium, and tin.

If hydroxides are the major component of a passive film, then the outermost surface of the interphase in contact with the solution is probably OH ions. This leaves open two possible reaction mechanisms for inhibitors, either by hydrogen bonding or acid/base interactions. The inhibition of steel by perfluoroalkylether aryl phosphine (FAP) is thought to act in this manner, as are many other organic compounds. However, little direct evident has been developed to support either of these mechanisms.

While the mechanism of metal hydroxide formation is certainly a known method of corrosion inhibition, the reaction of organic inhibitors by hydrogen bonding with OH^- or acid/base interaction would require a larger concentration of inhibitor for protection than is often observed. Therefore, inhibition involving either of these mechanisms is probably very limited and does not represent the primary means by which inhibitors work.

The mechanisms by which inhibitors protect metal surfaces are still largely unknown. However, with the advent and application of surface analysis techniques, the role of the inhibitor is becoming more apparent. As these mechanisms are unravelled, it is becoming increasingly evident that the corrosion film participates in one or more ways with the inhibitor to produce protection. Only in the limited cases of interface inhibition may there be little or no role of a corrosion film. Yet, the majority of systems that are inhibited are slightly acid to basic and encompass the presence and participation of interphase inhibition by films.

References

1. K. Juttner, W.J. Lorenz and F. Mansfeld, Corrosion/89, Paper No. 135, NACE, New Orleans, LA, 1989.
2. W. J. Lorenz and F. Mansfeld, *Electrochim. Acta* **1986,** *31,* 467.
3. C. Chakrabarty, M.M. Singh, and C.V. Agarwal, *Corrosion* **1983,** *39,* 481.
4. F.B. Growcock, Corrosion/88, Paper No. 338, NACE, St. Louis, MO, 1988.
5. G.A. Salensky, M.G. Cobb and D.S. Everhart, *Ind. Eng. Chem. Prod. Res. Dev.* **1985,** *25,* 133.
6. F.B. Growcock and V.R. Lopp, *Corrosion* **1988,** *44,* 248.
7. M.S. Vukasovich and J.P.G. Farr, *Mat. Perform* **1986,** *42,* 9.
8. I.L. Rozenfeld, *Corrosion Inhibitors,* McGraw-Hill, New York, 1981.
9. M. Seo and N. Sato, Corrosion/89, Paper No. 138, NACE, New Orleans, LA, 1989.
10. I.L. Rosenfeld, *Corrosion* **1981,** *37,* 371.
11. S. Szklarska-Smialowska, Corrosion/89, Paper No. 140, NACE, New Orleans, LA, 1989.
12. M. Sakashita and N. Sato, *Corrosion* **1979,** *35,* 351.
13. J. Schreifels, P. Labine, R. Gailey, S. Goewert, M. O'Brien, and S. Jost, Corrosion/88, Paper No. 22, NACE, St. Louis, MO, 1988.
14. A.G. Akimov, M.G. Astavjev, and I.L. Rosenfeld, *Zashchita Metallov* **1976,** *12,* 321.
15. B.D. Craig, *Corrosion* **1984,** *40,* 471.
16. E.C. French, R.L. Martin, and J.A. Dougherty, Corrosion/89, Paper No. 435, NACE, New Orleans, LA, 1989.
17. M. Oshawa and W. Suetaka, *Corrosion Sci.* **1979,** *19,* 709.
18. D. Thierry and C. Leygraf, *J. Electrochem. Soc.* **1985,** *132,* 1009.
19. C-T. Liu, J-C. Oung, and C-C. Su, Corrosion/88, Paper No. 358, NACE, St Louis, MO, 1988.
20. D.R. Arnott, B.R.W. Hinton, and N.E. Ryan, Corrosion/86, Paper No. 197, NACE, Houston, TX, 1986.
21. T.N. Wittberg, C.A. Svisco, and W.E. Moddeman, *Corrosion* **1980,** *36,* 517.

5

Films and Pitting Corrosion

Introduction

Pitting is one of the most common yet insidious forms of corrosion attack. It is highly unpredictable and difficult to model. The role of films in pitting corrosion is presented in three parts: pit initiation, pit propagation, and repassivation.

Pit Initiation

The key parameter most often used to describe resistance of a metal to pitting is the breakdown potential E_b or pitting potential E_p. For potentials more negative than E_p, the metal surface is passive, whereas above E_p (potentials more positive than E_p) pitting occurs. The more positive the breakdown potential, the more resistant a metal is to pitting. Figure 1 illustrates the E_p difference for various metals in a 3% NaCl solution at 30°C [1]. The mild steel does not display a region of passivity and therefore corrodes in a uniform fashion with no E_p. Nickel immediately initiates pits as it begins to passivate at +0.1 V. Type 316 stainless steel displays the highest pitting potential, about +0.5 V.

Four conditions considered necessary for pit initiation and growth are

☐ The breakdown potential must be exceeded.

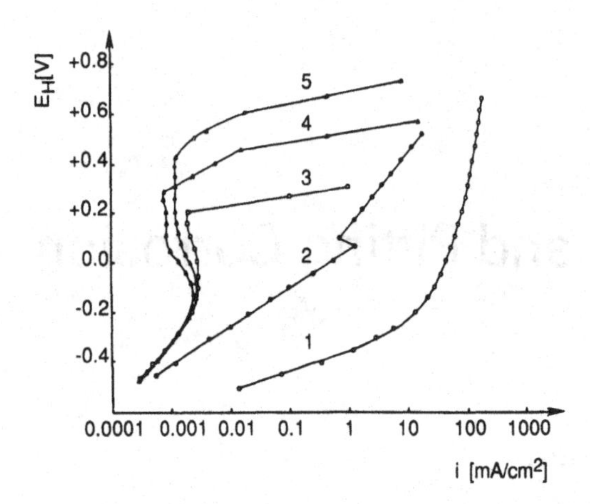

Figure 1. Slow scan anodic polarization curves (50 mV/3 min) in 3% NaCl solution at 30°C for (1) mild steel, (2) nickel, (3) 430 stainless steel, (4) 304 stainless steel, and (5) 316 stainless steel.

- ☐ The presence of an aggressive ion or species.
- ☐ An induction period prior to pit formation.
- ☐ The breakdown of the film at localized sites.

Recently it has been found that metastable pitting occurs below E_p and that pit initiation may occur below E_p. Stable pit growth is observed for potentials above E_p.

Several models have been put forth to describe the above considerations. All of them deal with local attack of the passive film. In the adsorption models, halide ions, for example, adsorb to the film at the film/solution interface adjacent to a cation in the film. In the presence of an adjoining anion vacancy, a metastable complex is formed between the cation and the chloride, resulting in immediate dissolution of the cation from the film. Since the electric field is essentially constant, it will promote the transport of another cation through the film to the film/solution interface. However, there being no additional oxide present for binding, the new cation is complexed and dissolution continues unabated.

In a similar vein, it has been proposed that aggressive ions, such as chlorides, might compete with oxygen for certain sites at the film/solution interface so that localized breakdown of the film occurs. In reality, it is difficult to determine whether site competition at the first monolayer of atoms of the film surface is the mechanism or if chloride ions are absorbed several monolayers into the film. In either case, the metal-halide bond apparently leads to enhanced dissolution compared to metal-oxide bonding. This mechanism has been proposed for the degradation of Al_2O_3 films on aluminum exposed to basic solutions containing chlorides [2]. The reaction sequence is as follows:

$$Al(OH)_3 = Al(OH)_2{}^+ + OH^-$$
$$Al(OH)_2{}^+ + Cl \rightarrow Al(OH)_2Cl$$

The product $Al(OH)_2Cl$ is a soluble salt that will then dissolve in solution.

The penetration models assume that film breakdown does not occur until the aggressive ion migrates or penetrates completely through the film to the metal/film interface. This penetration has been suggested to take place either by enhanced migration through pores in the film or by penetration through pore-free films. A variation on the latter concept considers the hydrated nature of oxide films and the interaction of chlorides with the water of hydration [3]. The chloride ions migrating through the film displace water molecules or hydroxyl ions, since the chloride ion is a stronger Lewis base than either H_2O or OH^-.

Another model suggests the preferential formation of a salt film in regions where the oxide film has broken down. This salt film is nonprotective and allows accelerated localized attack at these locations versus locations where the oxide film is more protective.

A statistical approach for modeling pitting corrosion has produced good correlation with experimental data. While the stochastic theory of pitting corrosion has been successful in describing such events as the dependence of pit initiation on potential, voltage sweep rate, and incubation time to initiate pitting, it is not a mechanistic model. That is, the stochastic model utilizes electrochemical parameters such as potential, current, time, etc., macroscopic variables which depend on film characteristics of an atomistic and electronic nature. Therefore, stochastic models in themselves are not useful in discussions on the specific role of films in pitting corrosion. However, combining specific attributes of the film with statistical analysis has proven quite successful in predicting film breakdown under anodic polarization [4].

The point defect model of pit initiation is illustrated with the aid of Fig. 2 [4]. During film growth, cation vacancies are generated at the film/solution interface in response to the potential drop across this interface. As a result of the concentration gradient, cation vacancies will diffuse and migrate in response to the electric field toward the metal/film interface. Similarly anion vacancies are generated at the metal/film interface and consumed at the film/solution interface. Figure 2 represents a stable film with movements of cation and anion vacancies in opposite directions but not influenced by aggressive ions. Figure 3 [4] presents a schematic view of the enhanced cation vacancy flux generated by the absorption of aggressive ions such as Cl^- into anion vacancies at the film/solution interface. If more cation vacancies arrive at the metal/film interface than can be consumed, a condensation (void) of vacancies will occur after some critical concentration is achieved, producing film breakdown. Of course, this model cannot predict if repassivation or future pit propagation will occur, it only describes the film stability/instability conditions that may initiate film breakdown. This model correctly predicts the benefit of increasing pH on pit induction time and the detrimental effect of increasing halide activity on decreasing pit induction time.

Passive film breakdown and pit initiation are not strongly dependent on film thickness. Rather, particular characteristics of the film, such as composition, semiconductive character (*p*- or *n*-type), and electric field strength, are more important. The kinetics of film growth and the degree of protectiveness are related to the presence or

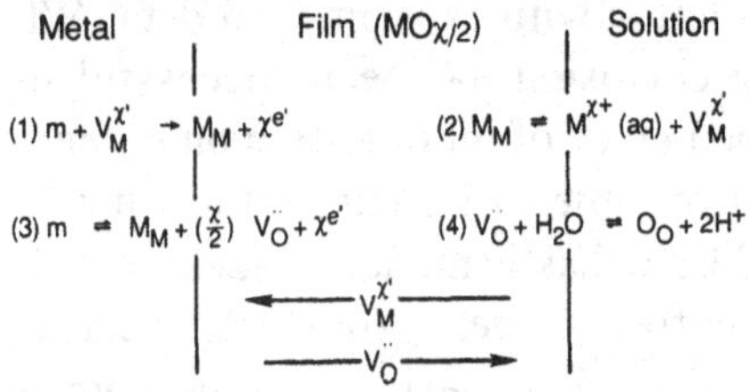

Figure 2. Schematic of physiochemical processes that occur in a passive film by the point defect model. O_O, oxygen ion in anion site; V_M^{ψ}, cation vacancy and $V_O^{\bullet\bullet}$, anion vacancy.

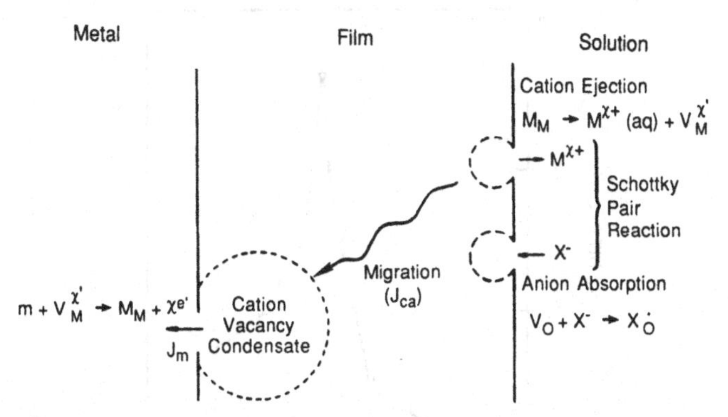

Figure 3. Enhancement of cation vacancy generation and condensation during breakdown of the passive film by the point defect model.

absence of chlorides in solution. For the case of steels and stainless steels, film composition is often unaffected. Therefore, electronic characteristics of the film and the easy migration paths provided by water incorporated in the film may control pit initiation and, eventually, propagation. Aggressive anions adsorbed on the film surface create a concentration gradient that exacerbates diffusion of these ions into the film due to the concentration gradient and the electric field of the film. Thus, transport of chlorides along water paths into the film in combination with easy movement of cations away from the metal will generate hydrolysis, reducing the local pH and causing dissolution of the film. Almost all aqueous-formed films will have some incorporated water, yet pitting only occurs above a specific pitting potential. Thus, the presence of water in the film is not sufficient in itself to initiate pitting, but a specific potential is also required that depresses the pH sufficiently along these water paths for pitting to proceed.

The specific characteristics of the corrosion film in regard to pitting resistance are demonstrated in Fig. 4 for type 304L stainless steel [5]. Samples preoxidized in dry air at different temperatures displayed completely different resistance to the degree of pit nucleation with a maximum in pit initiation observed for the sample oxidized at 300°C. This effect was found to be unrelated to film thickness, but by photopotential techniques it was demonstrated to be the result of *n*-type conductivity of the oxide film formed at 300°C.

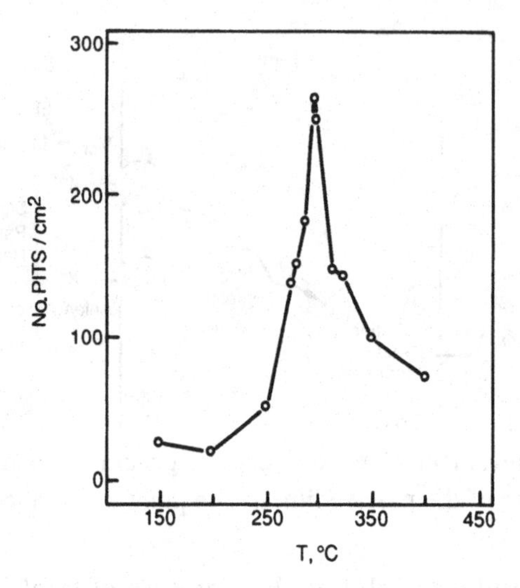

Figure 4. Pit density as a function of preoxidation temperature in dry air for type 304 stainless steel.

In fact, knowledge of the anodic transfer coefficient value, α_A, and the relative value of anodic transfer coefficient to cathodic transfer coefficient, α_c, have been used to relate the propensity for pit nucleation (Fig. 5) [5]. If the value of α_A is small and $\alpha_A \leq \alpha_c$, the film has *n*-type conductivity and pitting is enhanced. However, if α_A is large or exceeds some value specific to the film/environment system and $\alpha_A > \alpha_c$, *p*-type behavior is observed. Thus in a general sense,

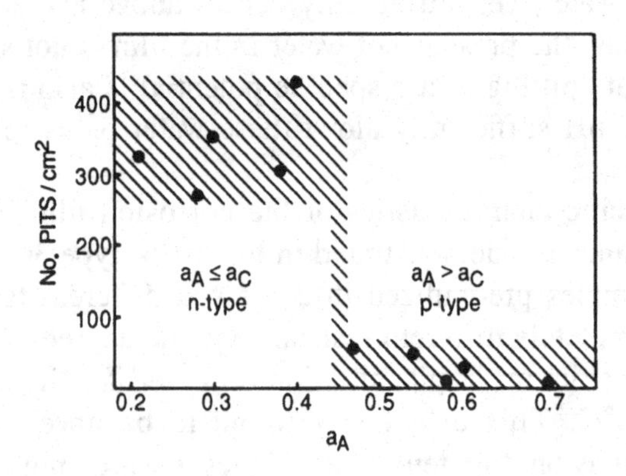

Figure 5. Dependency of pit density on anodic transfer coefficients.

films that possess anionic conductivity (*n*-type) will be less corrosion resistant than films that are primarily cationic conductors (*p*-type). Explained another way, *n*-type oxides are nonstoichiometric due to oxygen vacancies which enhance anion mobility and, therefore, chloride ions pass more easily through the film than do cations. Thus *p*-type semiconductor films should be more pitting resistant, all other factors remaining equal.

Similarly, if the induction time for pit nucleation is plotted as a function of chloride concentration and temperature, a linear relationship is obtained [6] (Fig. 6). Thus reduction in pitting resistance with increasing temperature can be explained by changes in film thickness, composition, and/or porosity. As already described, the effect from changes in film thickness are relatively small. Increasing porosity with increasing temperature is certainly possible, and over large temperature ranges it can be expected. However, as demonstrated in Fig. 2, the ion selectivity of the film also changes with increasing temperature from *p*-type at ambient temperature to *n*-type at

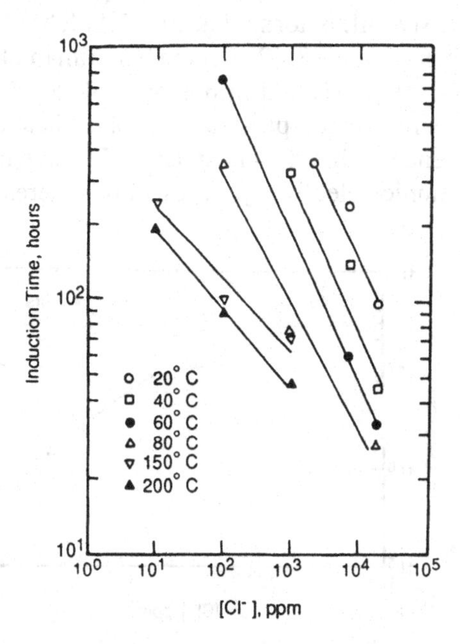

Figure 6. Induction time for pit initiation versus chloride concentration at different temperatures.

300°C. Therefore, pit nucleation in stainless steels is probably a complex relationship between the specific semiconductive properties of the film, the film thickness, and the porosity. Some of these factors may be at work, for example, in Fig. 7 [6]. The charge transfer resistance of the oxide on type 304 stainless steel is independent of chloride concentration at 40°C until a threshold concentration of approximately 8000 ppm Cl^-. At higher chloride concentration, the charge transfer resistance decreased, signaling a loss in corrosion protection by the film, yet no pitting was observed. This behavior implies either structural or compositional changes in the film have occurred that significantly alter the protectiveness of the film.

The ion selectivity of passive films formed on alloys other than stainless steel has also been observed. For instance, passive films formed on iron in borate and phosphate solutions exhibit enhanced pitting resistance to chlorides when the film is cation-selective, but they exhibit susceptibility when the film is anion-selective.

Figure 8 [7] illustrates the effect of ion selectivity on the transition time τ from pit nucleation to pit propagation. The thick line in Fig. 8 is for passive films formed at pH 11.5. The thin lines are for passive films formed at pH 8.42. At constant film thickness, τ is small in all cases for iron passivated in borate at pH 8.42 and exposed to chlorides, compared to iron passivated in phosphate at the same pH. The only difference is that the phosphate solution stimulates the formation of a cationic-selective (p-type) film, whereas borate stimu-

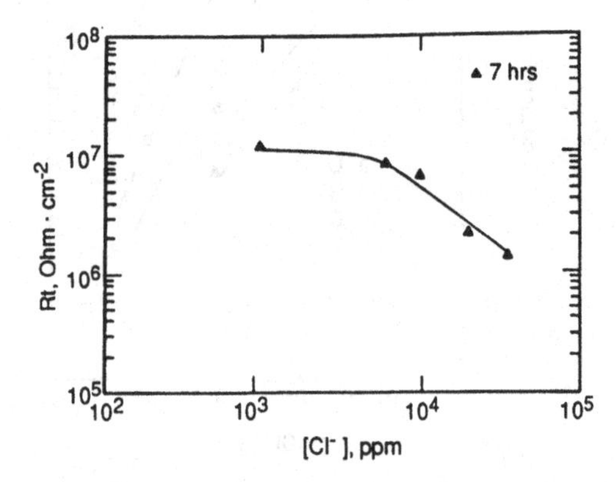

Figure 7. Charge transfer resistance, R_t, versus chloride concentration at 40°C and 7 hours exposure.

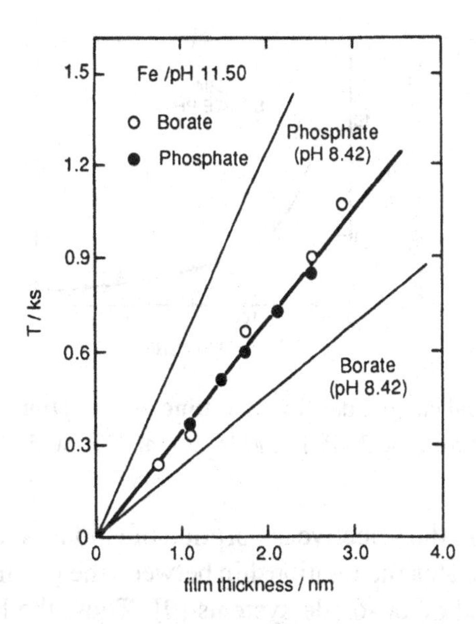

Figure 8. Effect of film thickness on τ for the passive films formed on iron at various potentials in borate and phosphate solutions at pH 11.5.

lates formation of anion-selective films. Isoselectivity or equilibrium for this system is pH 10.3. Thus the curve for pH 11.5 is slightly cation-selective and encompasses both borate and phosphate films. Therefore, by stimulating both *p*-type film conduction and increasing film thickness, the time required to begin pit propagation can be substantially delayed, thereby enhancing pitting resistance of the alloy.

The smooth transition from *p*-type semiconductive behavior to *n*-type is demonstrated in Fig. 9 [8]. Here type 304L stainless steel was passivated in 0.5 *M* $NaSO_4$ + 0.1 *M* $K_3Fe(CN)_6$ + 0.1 *M* $K_3Fe(CN)_6$ at different temperatures to produce a preformed oxide. The critical current density for pitting in 1 *M* H_2SO_4 with 100 ppm chloride was observed to decrease with increasing oxide preformation temperature (i.e., increasing *n*-type conductivity).

The pitting potential of oxide-covered metals can be related to the surface charge of the oxide at the film/solution interface. When the pH of the solution is greater than the pH of zero charge (pH_{zc}), the film has a net negative surface charge that promotes the adsorption of cations. Conversely, when the pH is less than pH_{zc}, a net positive

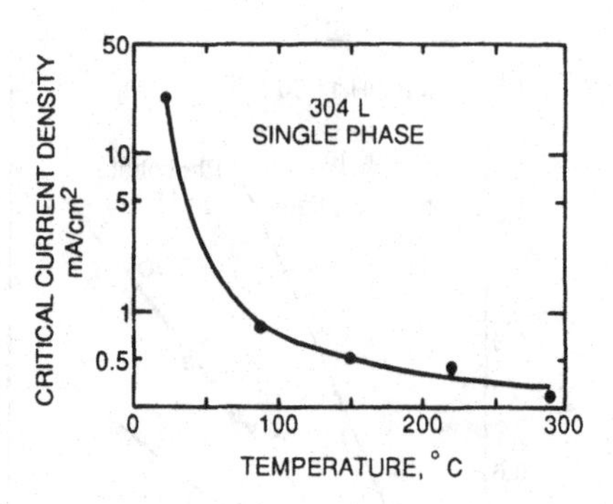

Figure 9. Critical current density for pitting as a function of film preformation temperature for type 304L in 1 *M* H_2SO_4 at 25°C with 100 ppm chloride.

charge results in the selective adsorption of anions, such as chlorides. Figure 10 illustrates the relationship between the pitting potential and pH_{zc} for several metal-oxide systems [9]. Thus, the lower the pH_{zc}, the more resistant a film-covered metal is to pitting attack. Moreover, any modification of the film to decrease its pH_{zc} would improve pitting resistance.

Doping aluminum oxide with 0.5 to 1 atomic percent TiO_2 has been shown to lower pH_{zc} to 7.5, while doping the same oxide with 0.5 to 1 atomic percent MgO increased the pH_{zc} to 9.6. Thus, modification of the film can substantially alter its resistance to pitting.

One method to accomplish oxide modification, and thereby enhance pitting resistance, is ion implantation. Figure 11 shows the results of implanting 4 atomic percent of ions (Mo, Si, Nb, Zr, and Al)

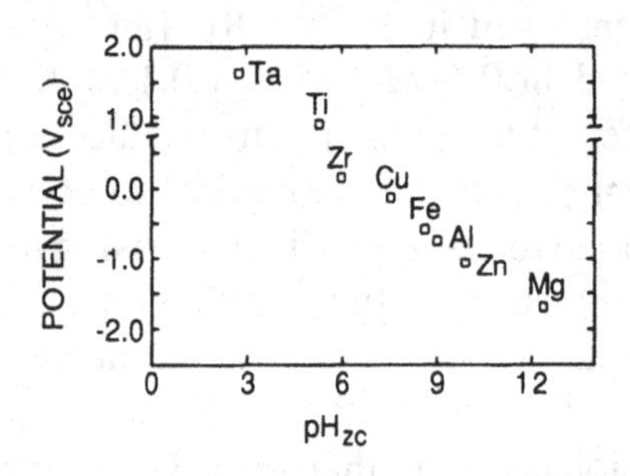

Figure 10. Pitting potentials of oxide covered metals versus the pH_{zc} of the oxide in 1 *M* NaCl.

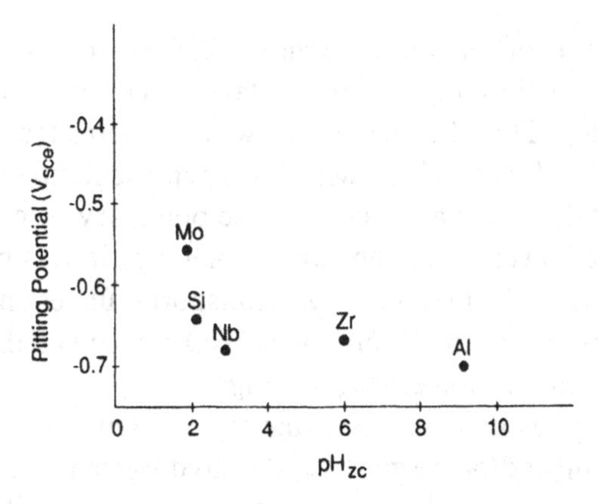

Figure 11. Effect of alloy doping with different metals on the pitting potential versus pH_{zc}.

in aluminum to investigate the resulting pitting potential in 0.1 *M* NaCl [9]. The benefit of certain ions to pitting resistance is obvious. Furthermore, the pitting potentials for unimplanted Al and Al with 12 a/o implanted Al were identical, demonstrating the effect is not due to some physical aspect of implantation but to an electrochemical effect.

Pit Growth

The transition from pit nucleation to propagation is gradual, yet not well understood. Several theories describe the growth of pits once they have been nucleated. One group of theories suggests the dissolution of metal ions in the pit and concomitant hydrolysis are a function of potential and pH. The potential at which the rate of production of hydrogen ions is equal to their rate of consumption by evolution is defined as a threshold below which passivation is dominant and above which repassivation does not occur [10].

Another model considers the rate-limiting step for pit development to be mass-transfer-controlled. The rate at which ions diffuse out of or into the pit entirely controls the growth rate [11].

Finally, there is considerable support for pit growth by salt film formation. Once saturation of a metallic salt is exceeded within the pit, precipitation of a salt film on the pit bottom occurs [12]. The salt film is formed by the cations of the metal and the aggressive ion (most often chloride). Thus further pit growth is controlled by transport across this film. Figure 12 is a schematic representation of a dual salt film on a metal and the associated phase boundary reactions [13]. In the unhydrated layer mass transport of metal cations is probably rate controlling, while the hydrated layer transport is under ohmic control. Repassivation can occur if sufficient H_2O molecules diffuse to the metal surface (a dilution effect, see Fig. 18).

Pickering has demonstrated that the accumulation of corrosion products in a pit and/or the generation of hydrogen bubbles can significantly limit the mass and charge transfer between the pit and the bulk solution. This produces a large ohmic potential drop in the solution from the pit cavity to the bulk solution (Fig. 13) [14]. The potential difference between the bulk electrolyte and the fluid inside a pit (slot) is minimal for shallow pits and for pits where no hydrogen bubbles accumulate. However, a large potential change occurs for deep pits that accumulate gas bubbles. Thus anodic reaction products and/or cathodic reaction products (gas) can significantly alter the potential inside a pit compared to the surface potential. However, the evolution of hydrogen from pits would be expected to increase the mass transfer rate, which could favor the second model of pit propagation, mass transfer control. These two models (salt film, hydrogen bubble) may be interdependent and are not easily distinguishable at this time.

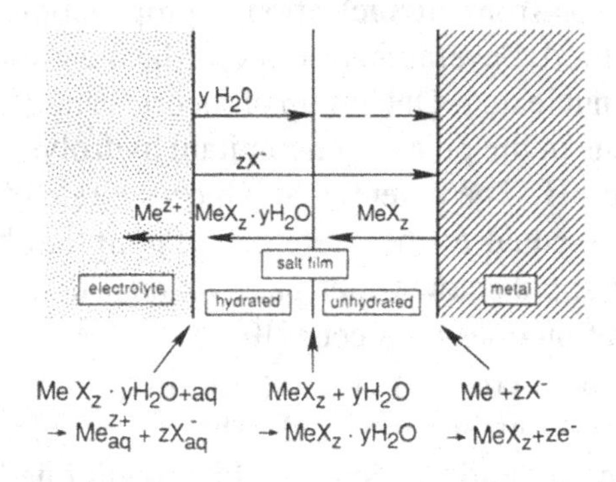

Figure 12. Schematic of the phase boundary reactions and transport across a dual salt film during anodic polarization in a chloride solution.

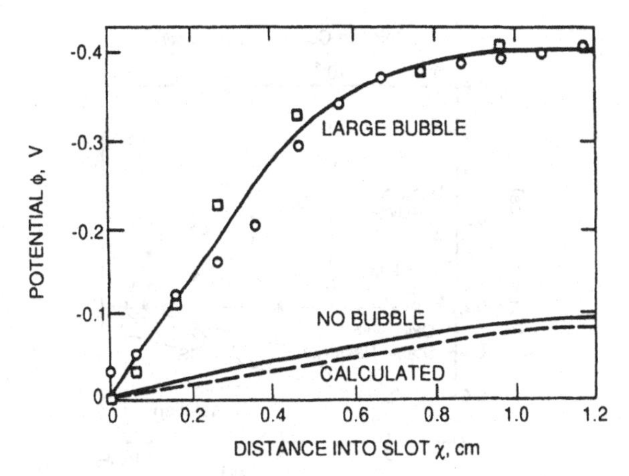

Figure 13. Potential change within an artificial pit in iron for the hydrogen evolution reaction at 52 A/m².

For many alloy systems, the early stage of pit growth is under ohmic control or a combination of ohmic and charge transfer control. However, once the pit becomes saturated in metallic ions, precipitation of a metal chloride can occur at the pit bottom. With the precipitation of a salt film on the pit bottom the controlling mechanism changes to diffusion control across this film [15]. The salt film governs the transport of metallic ions away from the metal and chloride ions through the film just as described for oxide films. Thus the electronic and ionic attributes of these films are of critical importance to the growth of pits. Yet essentially nothing is known about these important features of salt films.

Recently, the development of a stable salt film that controls pit growth has been quantitated to reflect transition from metastable pitting to stable pit growth. The critical potential for salt film formation, ϵ_{SF}, defines a specific potential below which active corrosion or passivity may occur and above which a stable salt film controls pit growth [16]. Figure 14 presents data for types 301, 302, and 304 stainless steels. It can be seen that the salt film formation potential is independent of pH and chloride ion concentration. However, the pitting potential (dashed line) decreases with increasing chloride until ϵ_{SF} and E_p meet at the highest chloride concentration. Below ϵ_{SF} the oxide is the controlling factor in the corrosion of a passivated metal,

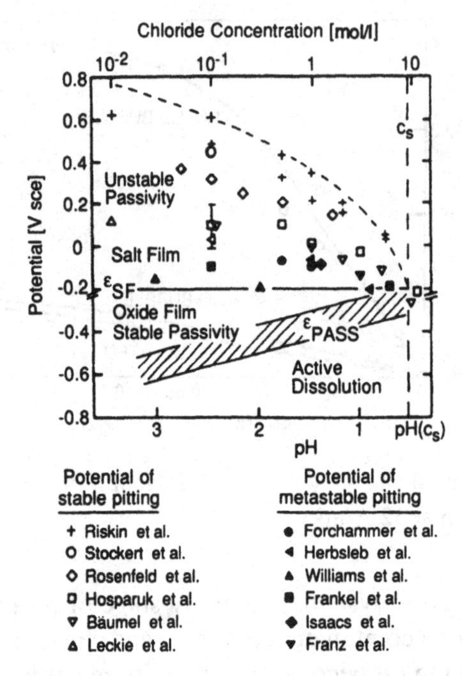

Figure 14. Schematic dependence of the passivation potential (ϵ_{pass}) on pH and the influence of chloride concentration on the potential of metastable and stable pitting. ϵ_{sf} is the salt film potential. These data are for types 301, 302, and 304 stainless steel.

and dissolution in pits that are present is under mixed ohmic/charge transfer control. Above ϵ_{SF} the salt film is more stable than the oxide, and dissolution is diffusion-controlled across the salt film.

The transition from ohmic/charge transfer control to diffusion control by the precipitation of a salt film is a function of potential and hydrodynamics. Low potentials, especially near the open-circuit potential, favor the ohmic/charge transfer mechanism as do small aspect ratio pits (large diameter), an agitated bulk solution and corrosion product flow out of the pits.

So numerous are the alloy systems that are susceptible to pitting in chloride solutions that most research on pitting is directed to the role of chlorides in passive film breakdown. Yet, this role is still not well understood. For example, it has been demonstrated that for pure iron passivated in borate buffered solution the adsorption of chlorides to the passive film takes approximately 1 minute, but absorption into the film takes about 1 hour [17]. The adsorption and absorption of chlorides were also found to be concentration- and potential-dependent. The absorption of chloride is related to a specific surface concentration of chloride and potential at breakdown (Fig. 15) [17]. Thus, at constant potential the absorption of chlorides into the passive film increases with increasing chloride content of the bulk solution, which is related to the increasing surface coverage of chlorides.

Contrary to this behavior, chlorides have not been observed to penetrate passive films formed on stainless steels even after substantial pit propagation has occurred. Rather, chlorides reside on or in the near surface of the film at the film/solution interface. However, much of the evidence for the absence of chloride penetration of passive films is based on techniques such as Auger that are not sensitive to very localized absorption of chlorides.

When molybdenum is added to an alloy for pitting resistance, it also does not become incorporated in the film. Therefore, there is no simple mechanism of Mo atoms diffusing into the film and nullifying the aggressiveness of the counterdiffusing chloride ion. It appears that Mo enhances the protectiveness of the oxide film on stainless steel by some mechanism not yet understood. If chlorides do not penetrate the film, the role of chlorides may be to selectively attack susceptible locations in the film. Thus, Mo apparently reduces the susceptibility of these sites to selective attack from chlorides.

In some cases, passive film breakdown appears to follow classi-

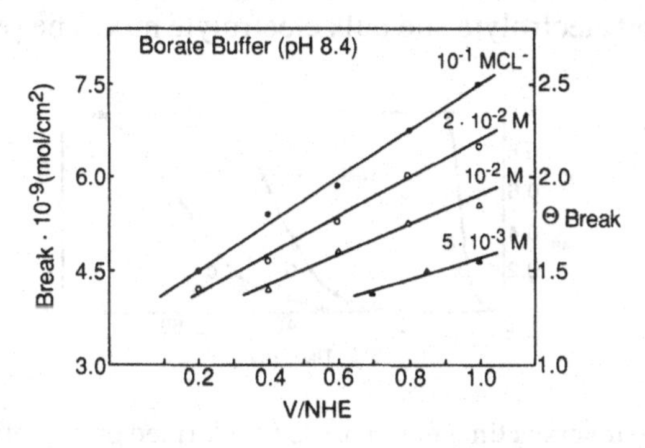

Figure 15. Total "surface" concentration of chloride at the breakdown potential for different chloride concentrations in solution.

cal nucleation and growth behavior. Figure 16 shows the relationship of the parameter $1-\alpha$ versus time for nickel in fluoride solutions. The parameter $1-\alpha$ is a measure of the extent of passivity breakdown. From Fig. 16 [18] it can be seen that breakdown begins after some incubation time and accelerates until complete breakdown occurs. Electron microscope examination of the nickel surfaces reveals nucleation of pits followed by lateral growth of the breakdown along the surface until the film is completely destroyed.

Repassivation

For many metals in a variety of solutions, there exists a potential below which repassivation may occur and pitting ceases. This potential is referred to as the *repassivation potential.* If the corrosion potential falls between the pitting potential and repassivation potential, growing pits will continue to propagate. However, a condition of metastability exists as pits initiate and attempt to reach stable propagation.

On stainless steels and nickel alloys, metastable pits grow with a passive film cover (Fig. 17) [19]. As long as this cover remains intact, metastable pitting progresses with transport out of and into the pit, regulated by diffusion through the passive film cover. This cover is not expected to be as protective as regions of the passive film that do not contain pits. Therefore, the cover is typically considered to have fine pores that enhance transport through the cover. Once the cover is punctured or destroyed as in Fig. 17, repassivation of the pits takes place as the pit electrolyte and bulk electrolyte mix. The repassivation

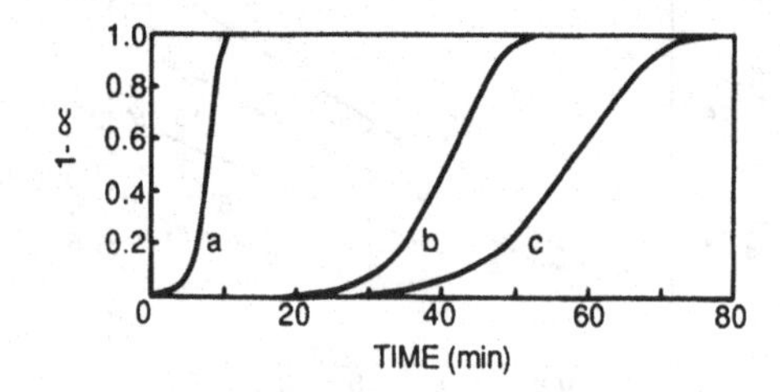

Figure 16. $1-\alpha$ versus time for an oxide film formed on nickel at 0.6 V for 1 hour in pH 3.0, sodium sulfate and broken down in pH 3.0 fluoride solution at (a) 0.4 V in 0.08 M F^-, (b) 0.1 V in 0.08 M F^-, and (c) at 0.4 V in 0.01 M F^-.

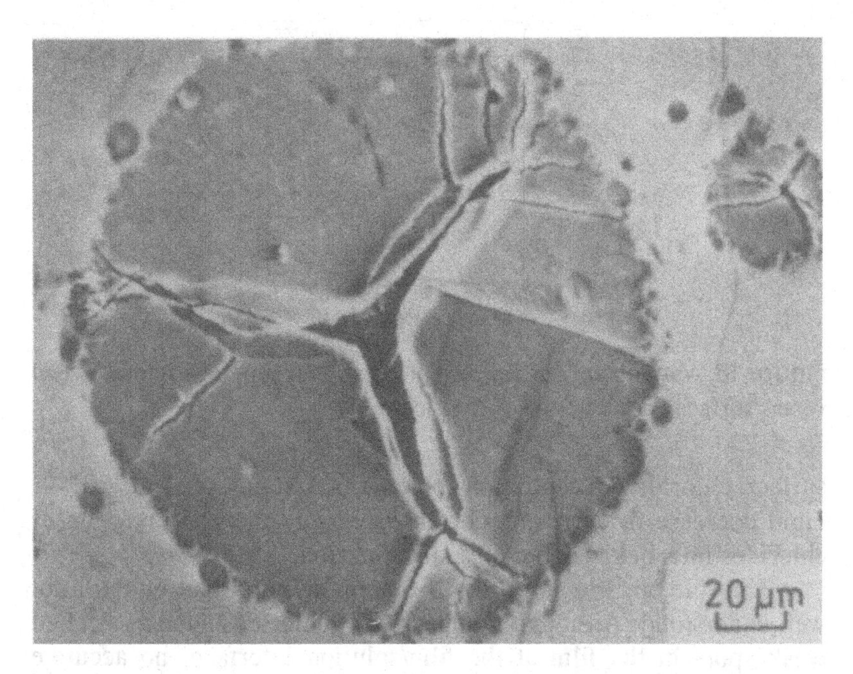

Figure 17. Ruptured pit cover on Ni–2.9% P in borate buffer with 0.1 *M* NaCl at 0.4 V.

only occurs if the potential is still below the pitting potential. Above the pitting potential stable pit growth continues even if the covering is damaged. Likewise, if the cover remains intact until a stable salt film develops on the pit bottom, the dissolution of the pit is stabilized by the ohmic potential drop across the salt film, and the cover is no longer necessary to ensure stable pit growth. Pit propagation will continue with or without the cover. The transition from metastable pitting to stable pit growth is accompanied by a current drop, which is probably related to the precipitation of the salt film. In other words, a supersaturated solution of salt in the pit will produce precipitation of a salt film, leaving a saturated solution that results in a current drop. Once salt film formation occurs, repassivation is unachievable and stable pit growth ensues.

Pit activity has also been found to depend on solution concentration in the pit. If the solution becomes sufficiently dilute, repassivation will occur. Figure 18 shows the repassivation dependence on

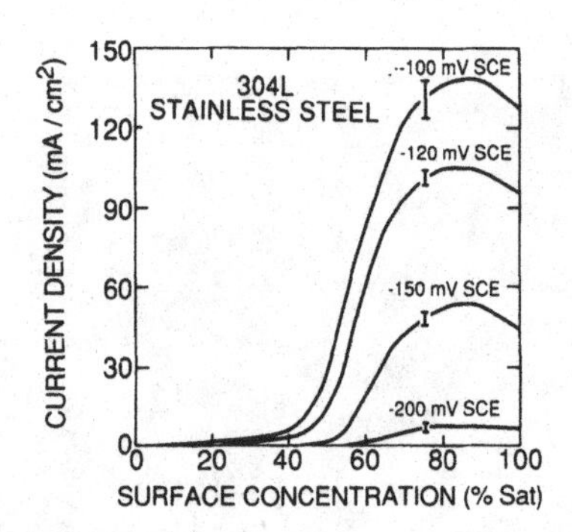

Figure 18. Variation of anodic current density for type 304 stainless steel versus surface concentration of chlorides.

surface chloride concentration for type 304 stainless steel [20]. A rapid decrease in corrosion current occurs as the concentration of chlorides falls below about 60% of saturation.

The role of films in pitting is extremely important but still not well understood. Although it appears that chlorides attack so-called weak spots in the film at the film/solution interface, no accurate description of what constitutes these weak spots, with the exception of the point defect model, has been presented. Nor have explanations for the beneficial roles of alloying elements such as Cr and Mo on pitting been established. Furthermore, much work remains to be carried out on the properties of pit covering and the defect structure of salt films at the bottom of pits.

References

1. Z. Szklarska-Smialowska, *Corrosion* **1971,** *27,* 223.
2. W.L. Archer, *Ind. Eng. Chem. Prod. Res. Dev* **1982,** 2, 670.
3. T.E. Pou, O.J. Murphy, V. Young, and J. O'M. Bockris, *J. Electrochem. Soc.* **1984,** *131,* 1243.
4. M. Urquidi-Macdonald and D.D. Macdonald, *J. Electrochem. Soc.* **1987,** *134,* 41.
5. G. Bianchi, A. Cerquetti, F. Mazza, and S. Torchio, *Localized Corrosion,* p. 399, NACE, 1974.
6. J.H. Wang, C.C. Su, and Z. Szklarska-Smialowska, *Corrosion* **1988,** *44,* 732.
7. R. Nishimura, M. Araki, and K. Kudo, *Corrosion* **1984,** *40,* 465.
8. P.E. Manning and D.J. Duquette, *Corrosion Sci.* **1980,** *20,* 597.
9. P.M. Natishan, E. McCafferty, and G.K. Hubler, *J. Electrochem. Soc.* **1986,** *133,* 1062.
10. J.R. Galvele, *Passivity of Metals,* p. 285, eds. R.P. Frankenthal and J. Kruger, The Electrochemical Society, Newark, NJ, 1978.

11. P. Roy and D.W. Fuerstenau, *Surface Sci.* **1972,** *30,* 487.
12. T.R. Beck, *Localized Corrosion,* p. 644, Ed. R.W. Staehle, NACE, Houston, TX, 1974.
13. M. Baumgartner and H. Kaesche, *Corrosion Sci.* **1989,** *29,* 363. (originally from T.R. Beck, *Electrochim. Acta* **1985,** *30,* 725.)
14. H.W. Pickering, *Corrosion* **1986,** *42,* 125.
15. F. Hunkeler and H. Bohni, *Corrosion* **1984,** *40,* 534.
16. F. Hunkeler, G.S. Frankel, and H. Bohni, *Corrosion* **1987,** *43,* 189.
17. V. Jovancicevic, J. O'M. Bockris, J.L. Carbajal, P. Zelenay, and T. Mizuno, *J. Electrochem. Soc.* **1986,** *133,* 2219.
18. B. MacDougall and M.J. Graham, *J. Electrochem. Soc.* **1983,** *132,* 2553.
19. J. Flis and D.J. Duquette, *Corrosion* **1985,** *41,* 700.
20. H.S. Isaacs, *Corrosion Sci.* **1989,** *29,* 313.

6

Role of Films in Stress Corrosion Cracking and Hydrogen Embrittlement

Introduction and Definition of Stress Corrosion Cracking

Stress corrosion cracking (SCC) is a term often used to describe the catastrophic fracturing of materials under the simultaneous influence of stress and exposure to an aggressive environment. This type of failure or fracturing is also referred to as *environmentally assisted cracking,* which emphasizes the important contribution of the environment on SCC. However, environmental cracking is also commonly used to describe all forms of cracking that are assisted by their environment, including liquid and solid metal embrittlement, hydrogen damage, and corrosion fatigue. To avoid confusion and the vagueness of the term *environmental cracking*, we use *stress corrosion cracking* in the classical sense to strictly apply to that form of cracking that incorporates a tensile stress with anodic dissolution to induce fracture. Furthermore, SCC is sometimes used to include classical *hydrogen embrittlement* (HE). While there are certain systems where SCC does appear to be dominated or assisted by HE, the majority of systems that display SCC cannot be simply classified as HE. Therefore, HE or, more correctly, *hydrogen damage* (HD) will be discussed as a separate mechanism.

Crack Initiation and Propagation

In the classical description of SCC and for most other forms of environmentally assisted fracture, cracking is divided into two separate events: initiation and propagation. In reality, there is probably no such convenient distinction, but for purposes of modeling and research the distinction is useful.

Figure 1 is a typical illustration of SCC behavior, also referred to as *static fatigue*. The total time to failure, t_f, consists of a crack initiation component and a crack propagation component. The relative time spent in initiation compared to propagation depends on many factors and continues to be at the core of SCC research. In some alloy-environment systems, such as high-strength steels in seawater, crack initiation may be the rate-controlling step since propagation occurs quite rapidly once a crack has initiated. However, in lower-strength alloys near the threshold for cracking, K_{ISCC}, cracks may initiate, but propagation is either imperceptibly slow or nonexistent.

Crack initiation is strongly dependent on many surface features, such as the presence of pits or crevices, surface finish, degree of cold work, microstructural features, inclusions, and weld profiles. Any of these features can significantly reduce the time for crack initiation, thereby reducing the total time to failure. However, there are many alloy/environment systems for which pitting, crevices, etc., occur, but SCC does not result. One of the major difficulties in pursuing an understanding of the fracture initiation stage of SCC is the develop-

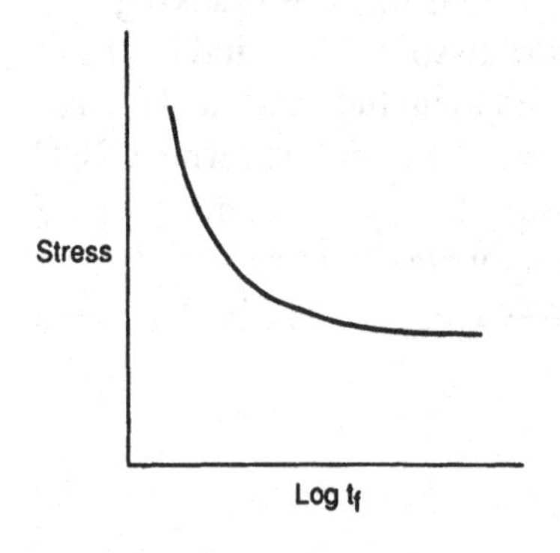

Figure 1. Classic illustration of SCC behavior showing the time to failure (t_f) of a metal as a function of either applied stress or yield strength.

ment of an adequate definition for the transition of a pit or crevice to a crack.

As might be anticipated, pit or crevice geometry would be an important factor in determining the probability for SCC initiation. Generally, an aspect ratio, pit depth to width, of 1 reflects general or uniform corrosion, while ratios on the order of 1000 or greater are typical of growing stress corrosion cracks [1]. Aspect ratios in excess of 10 are considered potential sites for crack initiation. However, there are numerous exceptions to these general guidelines, and the particular aspect ratio for crack initiation from such a site will depend on the stress state ahead of the crack tip and the fracture toughness of the bulk alloy. For instance, nonpropagating cracks are evident at the root of surface trenches in a 2% Ni steel exposed to H_2S near the K_{th} for this alloy (Fig. 2). The aspect ratio approach to crack initiation emphasizes the mechanical or stress component of SCC, but electrochemical factors may be just as important, if not more so, in many systems.

In order for a pit or crevice to proceed to the stage of crack initiation, the electrochemical reactions at the crack tip must progress at a faster rate than the dissolution processes taking place on the metal surface or along the walls of the pit. If the latter occurs, pit geometry necessary for crack initiation is not maintained and general corrosion or pitting attack continues. In order to maintain sufficient pit geometry for crack initiation, the bottom of the pit (crack tip) must remain electrochemically active while the pit sides move from active to passive behavior. This requirement of activity at the pit tip for crack initiation, especially from surface-originated artifacts such as pitting, has been the focus of significant research on pit initiation and propagation, a precursor to SCC.

Sharp pits (high aspect ratio) are often considered to create a suitable site for SCC primarily by the stress concentration developed. While this may be a contributing factor, it is apparently not a predominant one. Since pits that initiate SCC need to possess both an active tip of the pit and inactive sidewalls, it is obvious the potential of the pit and composition of the electrolyte must vary significantly from the tip out to the surface where bulk solution conditions exist. As the notch (crack) acuity increases, this difference between crack tip solution chemistry and bulk chemistry will be accentuated due to the restricted access of solution to the tip. Furthermore, the potential at the crack tip would also be expected to vary as a result of the solution composition gradient. These localized differences in electrochemical potential and chemistry within a pit are discussed later.

One other contribution to crack propagation in sharp cracks is the wedging open of the crack from oxides filling the crack mouth. The major contribution from oxide wedging is a resulting residual stress which has been found to predominate closer to the crack opening than to the crack tip. As such, the wedging effect may be most important for small crack sizes (early crack growth) and during crack initiation.

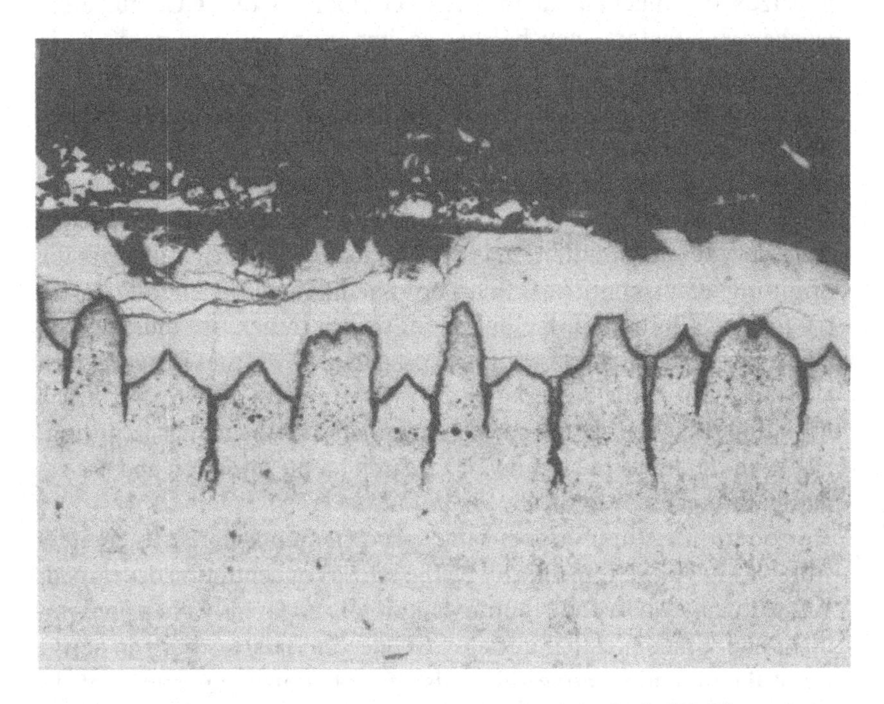

Figure 2. Ferrous-sulfide-filled trenches and cracks in 2.0% Ni steel exposed to aqueous solution of 5% NaCl and 0.5% acetic acid saturated with hydrogen sulfide.

Current Theories of SCC

It is generally held that SCC occurs in metal-environment systems that produce a film or scale on the metal surface. Yet not all environments that form films on metals create the necessary conditions for SCC. In fact, only certain alloy-environment combinations produce necessary and sufficient conditions for SCC, albeit the list of possible combinations continues to grow rapidly in step with the high level of research activity in the field of SCC research.

To gain a better perspective on the role of surface films on SCC, we briefly comment on the more prominent theories of SCC. Three general categories are used to classify SCC models:

a. Dissolution or active path corrosion
b. Adsorption induced fracture
c. The film rupture mechanism

The *active path corrosion* (APC) mechanism was one of the earliest models presented to describe SCC. It is most successful in describing the preferential SCC along grain boundaries that contain precipitates that are anodic to the surrounding grain matrix, for example, the precipitation of beta phase (Mg_2Al_3) in Al/Mg alloys. However, beta must be present in an essentially continuous phase to create conditions for APC in these alloys. The same situation occurs in the Al/Cu and Al/Cu/Mg alloys.

In a similar fashion, other phases on the boundaries may be cathodic to the grain matrix yet create a denuded zone adjacent to the boundary that is anodic to the boundary, resulting in the same SCC susceptibility. The classic example of this circumstance is precipitation of chromium carbides at grain boundaries in austenitic stainless steels—sensitization. The active path in this case is adjacent to the boundary since the precipitates will be cathodic to the denuded region which is low in Cr.

APC has also been applied to intergranular cracking along grain boundaries that have enhanced concentration of impurities due to segregation. While pure iron is generally resistant to SCC, especially in nitrates, steels containing various impurity elements (i.e., P, Sn, Sb, etc.) become susceptible to cracking. The magnitude of these effects is shown in Fig. 3 [2]. Phosphorus and sulfur are seen to be more deleterious than Zn and As. However, this relative ranking is a function of the specific environment. Time to fracture in ammonium nitrate is significantly reduced by the segregation of certain elements to prior austenite grain boundaries. However, the pure steel (0.1% C) was essentially resistant to cracking.

The development of surface science techniques to study surfaces (ESCA, Auger, SIMS) has allowed quantitative investigation of grain boundary segregation and further study of the role of segregation in SCC. Even before these techniques were developed, the grain boundary enrichment of alloys could be predicted by classical segregation theory based on atomic size and thermodynamic considerations. Figure 4 illustrates the significant enrichment of boundaries by solutes that may occur in various alloys [3].

Phosphorous is a typical example of an impurity in steel that at sufficient concentrations can embrittle an alloy without the contribution of a corrosive environment. More recently, the role of P on SCC and hydrogen-assisted cracking has been found to be quite important [4]. In order to study the contribution of P segregation on APC, an evaluation must be done independent of any film contribution. By

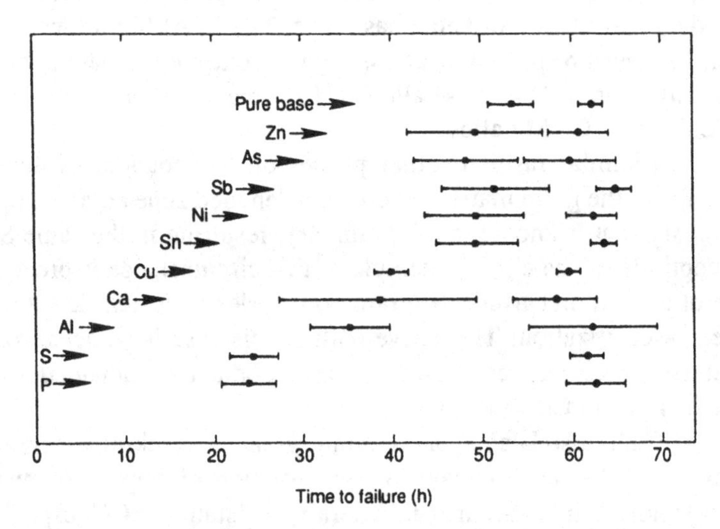

Figure 3. Fracture times of 11 alloys stress corroded at a constant strain rate. ●, tests in paraffin; ■ , tests in NH_4NO_3 solution.

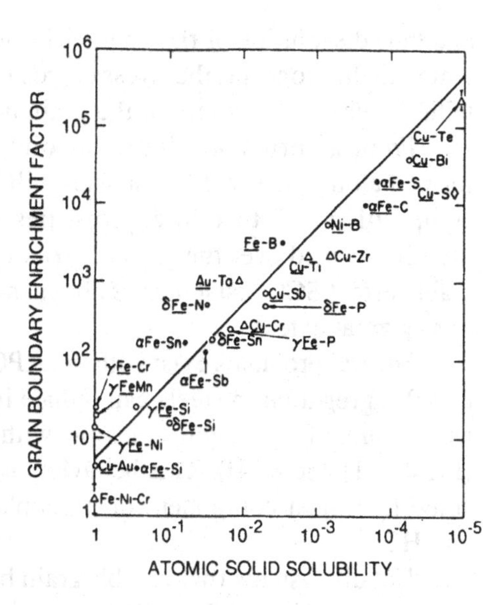

Figure 4. Correlation of the grain boundary enrichment factors with atomic solid solubility at the temperature of the experiments. Open circles are obtained by the interface energy approach; filled circles by Auger electron spectroscopy; open triangles by self-diffusion in the lattice and the grain boundaries; open diamond by sulfur diffusion.

cathodically polarizing the sample to reduce any oxides present and then applying a galvanostatic pulse, a corrected potential E_{p*} can be developed that corresponds to the net applied galvanostatic pulse current density i_a and the two parameters plotted (Fig. 5) [5]. Thus, the film-free anodic dissolution kinetics for a commercial alloy can be compared to the intermetallic compounds of Fe_3Sn and Fe_3P, which would simulate the grain boundary concentration of these impurities. It can be seen that the boundaries would be anodic to the bulk alloy,

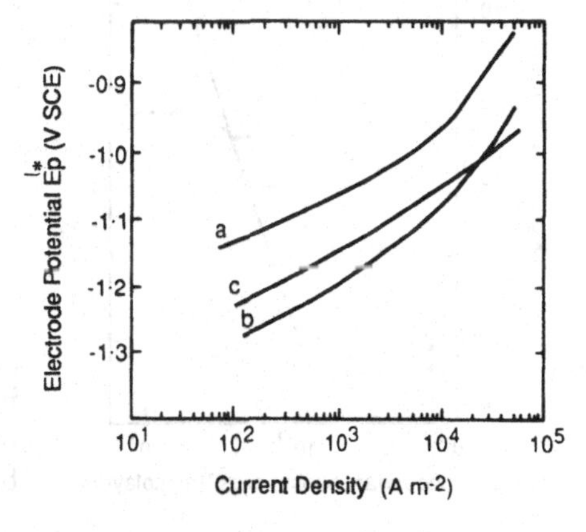

Figure 5. Film-free anodic polarization curves for (a) Fe/3%Ni/0.09%Sn, (b) Fe_3Sn, and (c) Fe_3P in 5 *M* NaOH.

and thus dissolution of the boundaries would be more rapid. Consistent with this concept, the stress corrosion crack growth rate increases with increasing P content of the grain boundary region (Fig. 6) [6].

From a corrosion film standpoint, it is expected that the film formed along the grain boundaries will have a significantly different composition and, therefore, properties different from that portion of the film formed over the grain matrices. This condition could appreciably effect SCC resistance of alloys, but it has not been investigated to any great extent.

Several problems exist with the APC mechanism. For instance, S and P segregation in steels can induce intergranular fracture entirely as a result of hydrogen cracking with no contribution from APC (Figs. 7 [1] and 8 [4]). This behavior suggests that APC may involve some hydrogen component or a coupled mechanism between APC and HE.

Not all systems that exhibit grain boundary segregation are susceptible to SCC, and those that are susceptible are only susceptible to specific environments, not every one that promotes SCC. Furthermore, susceptibility to intergranular corrosion is not always a precursor for SCC. Moreover, the important contribution of stress is minimized in the APC mechanism for SCC.

Of course, one of the obvious shortcomings of the APC approach

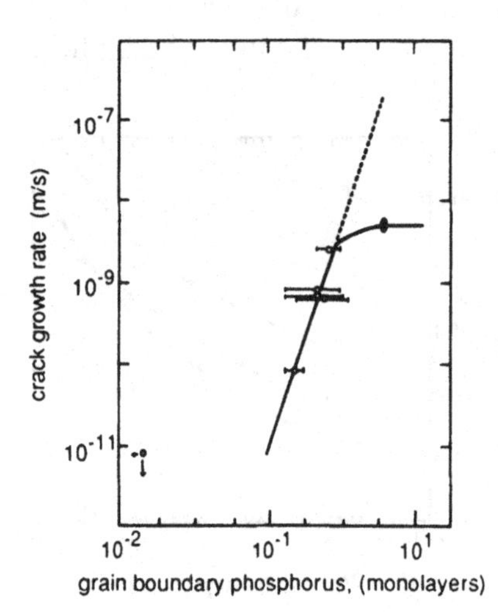

Figure 6. Stress corrosion crack growth rate (for 3 Cr–0.5 Mo steel in 8 *M* NaOH at 115°C at −354 mV_{NHE}) as a function of grain boundary phosphorus coverage.

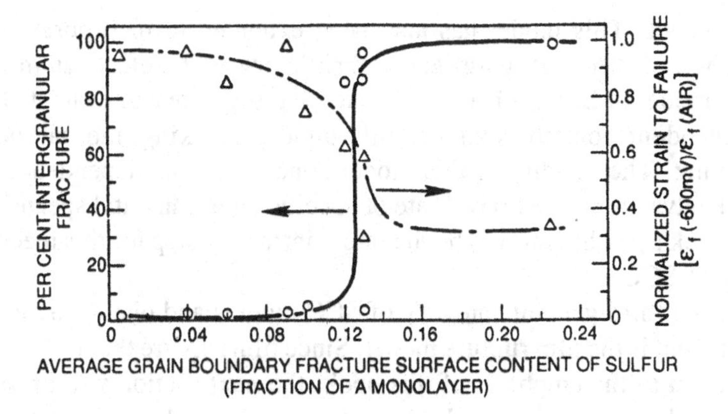

Figure 7. Percent intergranular fracture and the normalized strain to failure plotted as a function of sulfur content at the grain boundary for straining electrode tests at a cathodic potential of -600 mV (SCE).

is its dependence on intergranular attack which does not address the numerous systems for which SCC is transgranular.

The specificity of certain ions producing SCC in specific alloy systems led to the development of the specific ion or, more generally, the adsorption-induced fracture model for SCC. This model suggests that chemisorption of certain ions will enhance nucleation of dislocations at the crack tip, promoting brittle cleavage fracture over more ductile-type fracture modes typically observed for most alloys. Both intergranular and transgranular fractures have been explained with

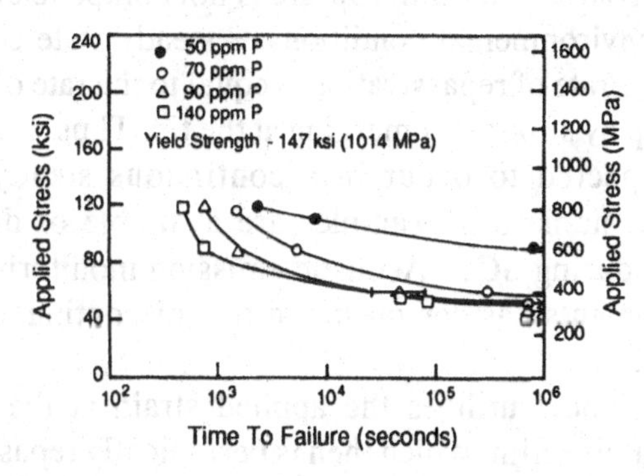

Figure 8. Applied stress versus time to failure in hydrogen for 4130 steels containing different levels of P.

this model. This model has also been extended to incorporate the decohesion theory of hydrogen embrittlement. Therefore, as in the APC model of SCC, it becomes difficult to separate and evaluate the independent contributions of truly anodic cracking and cathodic cracking. The validity of this model is based on the dependence of crack advance on the arrival rate of specific environmental species to the crack tip, which would be the rate-controlling step for this mechanism of SCC.

The third general category of SCC models and one of the most prominent is the film rupture model. Since film rupture theory is most pertinent to the emphasis of this book, the next section will present this mechanism in detail. However, this emphasis does not diminish the relative importance of contributions of the other models to SCC.

Film Rupture Model of SCC

There are many SCC models based on film rupture or that incorporate film breakdown in the SCC mechanism. Rather than address each of these individually, they are collectively discussed with emphasis on the major points that elucidate the role of films. In its most general sense, the film rupture model for SCC assumes localized breakdown of a passive or protective film that is not repassivated, allowing significant dissolution at the bare crack tip while the crack walls remain protected by the passive film.

If repassivation after film rupture is not complete due to electrochemical or environmental conditions, a steady-state condition may arise where the rate of repassivation is equal to the rate of film rupture at the crack tip by slip steps emerging at the tip. Thus, crack extension would be expected to occur in a continuous steady-state manner. However, numerous examples are available of discontinuous crack growth during SCC. Acoustic emission monitoring of SCC in many environments has demonstrated this discontinuous manner of crack growth.

Another model utilizes the applied strain at the crack tip to rupture the passive film, which then is periodically repassivated. This cycle continues producing discontinous crack growth [7] (Fig. 9). Of course, simply providing an electrochemically active crack tip does

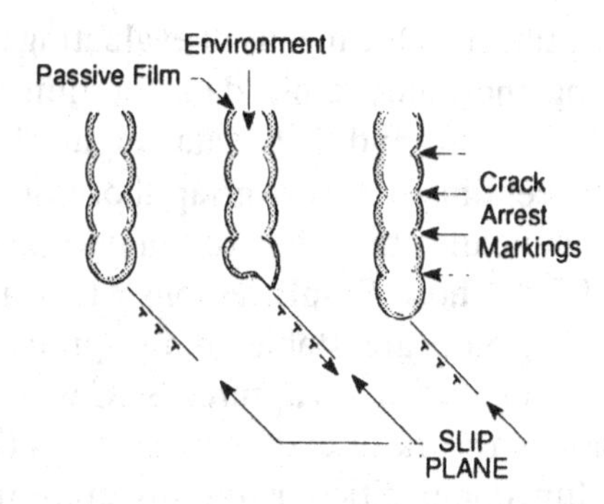

Figure 9. Schematic representation of crack propagation by the film rupture model.

not ensure a sharp crack tip, which is necessary for continued propagation. On the contrary, dissolution of the crack tip would be expected to produce tip blunting, resulting in crack arrest if no other factors are involved.

To address the need for sharp cracks and discontinuous crack growth, modification of the above models have considered the film or a thin dealloyed layer of the surface as the primary site for brittle fracture initiation. In this model, a crack originating in the oxide penetrates for some distance into the ductile alloy where the crack is arrested and blunted. The environment then produces the necessary brittle film at the crack tip and along the crack walls, and the oxide again ruptures. The cycle repeats itself until a critical crack size is achieved for unstable crack growth. Variations of this model have been used to demonstrate both intergranular and transgranular fracture, and direct observations of such events have been documented [8]. This model requires a thin brittle oxide that is coherent with the substrate material or a thin dealloyed layer on the surface.

Evidence of Films in SCC

Specific knowledge of the role of film composition, structure, and properties (electronic, ionic transport, etc.) associated with SCC is quite sparse; however, some attempts have been made in recent years to understand the relationship between some of these parameters and SCC susceptibility. One means of evaluating this relationship has been combining thermodynamic data for film formation from E_H-pH (Pourbaix) diagrams and SCC data. Figure 10 represents the regions for SCC susceptibility that overlap those of protective films that are thermodynamically stable for iron in phosphate, nitrate, carbonate, and water [1]. In those E_H-pH regions where a soluble species such as Fe^{3+}, Fe^{2+}, etc., are stable, if the protective carbonate, phosphate, nitrate, or oxide film is ruptured SCC will generally occur. A similar example from work in carbonate systems (Fig. 11) reflects the importance of film composition on the susceptibility or resistance to SCC [9]. Cracking was almost always associated with a black film that was predominantly Fe_3O_4 and was intergranular. The boundaries of the SCC region presumably represent specific oxide reactions, especially since outside the cracking regime no films were detected. These results emphasize the dependence of SCC on the presence and composition of surface films.

Other systems besides iron and steel have been examined using the Pourbaix approach. For instance, 70/30 brass in hydroxide, formate, acetate, and tartrate is shown to SCC under specific domains of stability for various chemical species (Fig. 12) [1]. Again, thermodynamically stable films apparently promote SCC (intergranular in this case).

While this approach to studying participation of corrosion films

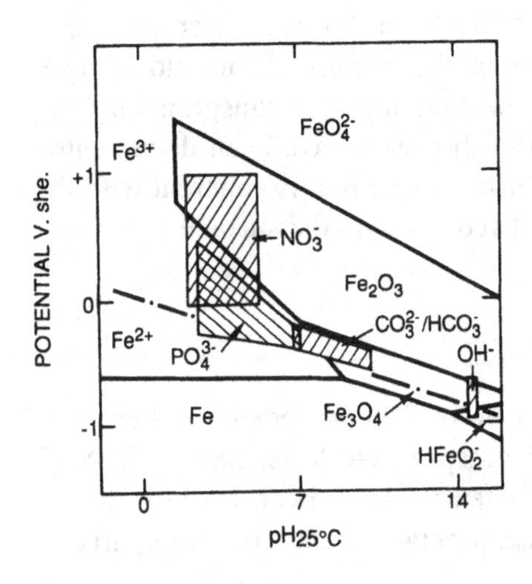

Figure 10. Relationship between E_H-pH conditions for severe cracking susceptibility of mild steel in various environments and the stability regions for solid and dissolved species.

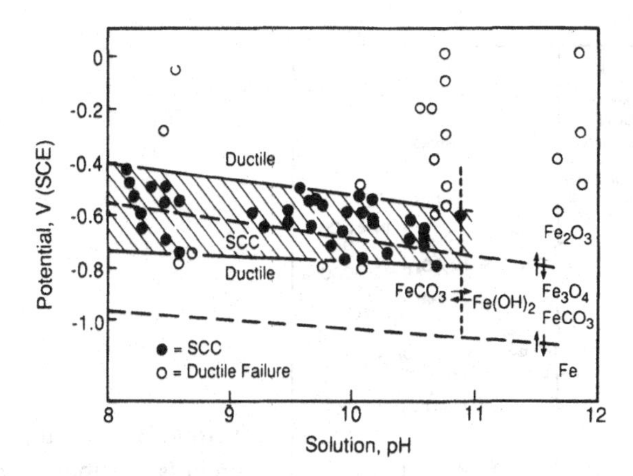

Figure 11. Potential-pH diagram showing stress corrosion cracking domain defined by slow strain rate tests in MEA solutions.

in SCC is useful, it has several limitations. Pourbaix diagrams have not been developed for most of the complex alloy systems currently under study, nor are they available for the range of temperatures necessary to fully evaluate SCC for many alloy-environment systems. More importantly, the crack tip potential, pH, and chemistry can be significantly different from the bulk solution properties. Figure 13

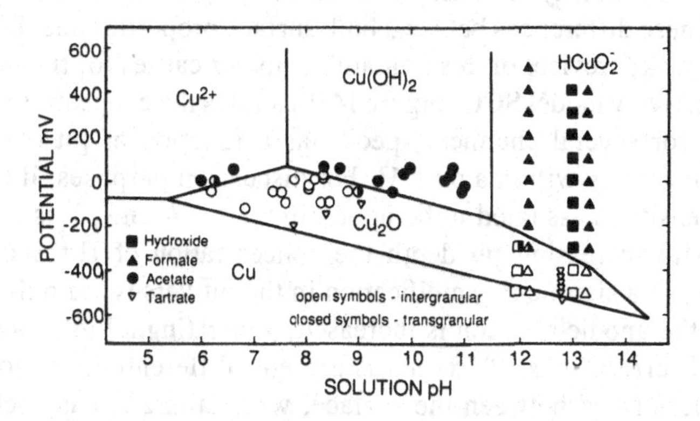

Figure 12. Potential-pH diagram showing the domains of failure mode for 70/30 brass in various solutions, compared to the calculated positions of various boundaries relating to the domains of stability of different chemical species.

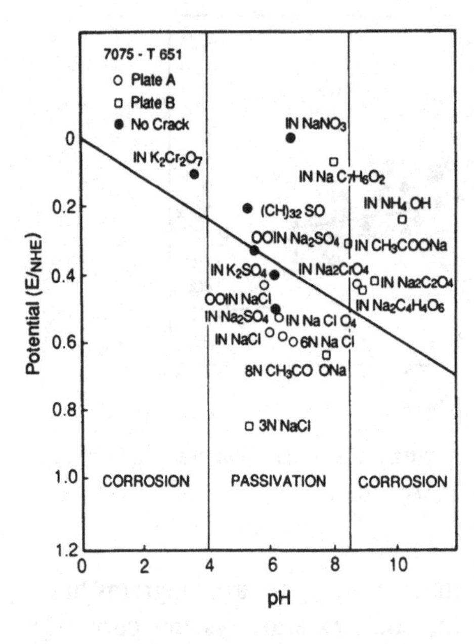

Figure 13. Cracking data (open symbols) superimposed on Pourbaix diagram. The filled circles refer to plate A. The diagonal line is for hydrogen reduction. The stable oxide phase (passivation) is hydrargillite.

reflects the discrepancies involved for SCC compared to Pourbaix diagrams [10]. Cracking in the passivation field, as well as the active corrosion fields, is evident, as is noncracking behavior in these regions. Therefore, the use of thermodynamic diagrams for predicting SCC susceptibility is limited and can be incorrectly applied.

These differences between bulk surface properties and those in a tight crack, crevice, or deep pit are the major causes for the inability to effectively model SCC. Figure 14 illustrates the concentration variations for several chemical species as a function of pit depth and current density within a pit [11]. For discussion purposes, if the current density is assumed to be approximately 1 A/cm^2, it is apparent that with increasing pit depth the concentration of H^+ increases, leading to a significant acidification in the pit versus the bulk. Likewise, the anodic reaction is increased, generating a high concentration of ferrous ions. If oxygen is present, differential aeration cells may be formed between the surface, where there is easy access to oxygen, and the crack tip with limited oxygen, further stimulating the advance of the pit bottom or crack tip. The crack tip/side wall system is thus seen to be a dynamic system rather than a thermodynamically stable environment. Therefore, the evaluation and study of SCC en-

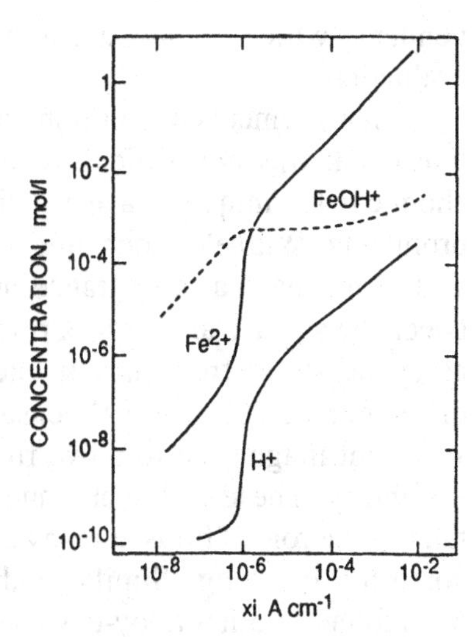

Figure 14. Concentration of Fe^{2+}, $FeOH^{+}$, and H^{+} as function of the product, depth x, and current density i, in unidirectional pit.

tails both the thermodynamics, as well as kinetics of a system and the realization that the crack tip environment may be significantly different from bulk solution properties.

More work probably has been done on crack tip environments for corrosion fatigue cracks than any other system. It has generally been observed for simulated cracks that when the external surface of a specimen is anodically polarized, the crack tip potential becomes increasingly negative as cyclic strain, temperature, and solution flow past the tip increase. These factors then induce a large potential drop from the crack mouth to the tip. If the external surface is cathodically polarized, a potential gradient is again developed, with the tip moving to more positive potentials. However, under these conditions, cyclic strain does not have an effect on potential distribution when the crack tip is in the cathodic regime because no surface films are present for rupture by these strains.

The magnitude of these potential drops and the potential distribution along the crack walls are governed by the dimensions of the crack, the ohmic drop within the crack, the local solution chemistry, reactions and reaction rates within the crack, solution conductivity, transport of species by diffusion and ionic migration, convection within the crack, the composition and properties of the films formed on the side walls, and the temperature. All of these factors are interdependent, which creates a complex environment for modeling and evaluation.

Measurements of solution composition, pH, and potential at actual crack tips are difficult to obtain. Recent advances in electrochemical techniques have provided preliminary results that look promising. With electrochemical impedance spectroscopy, combined with a mechanical admittance analog, the potential difference between the surface and a crack tip was found to depend on factors such as cyclic stress frequency in the case of fatigue (Fig. 15) [12]. A difference of 200 mV, as measured, is significant and could be of sufficient magnitude to move from a highly stable film to complete instability. These new techniques promise to provide additional details on the role of films in many aspects of corrosion, including stress corrosion cracking. Similar and greater potential differences have been found in other alloy-environment systems.

Critical potentials for SCC can be determined by using potentimetric methods combined with the thermodynamic necessity of simultaneous film formation and oxidation at the crack tip. Figure 16 shows critical potential ranges for intergranular and transgranular SCC for several alloys [1]. Two distinct ranges for critical potential are apparent in Fig. 16, both of which correspond to the transition from active to passive or passive to transpassive behavior. In both cases, the requirements for passive crack walls and active crack tip are met. In all other potential regions either complete passivation with rapid repassivation of film rupture sites occurs, or rapid uniform dissolution is dominant.

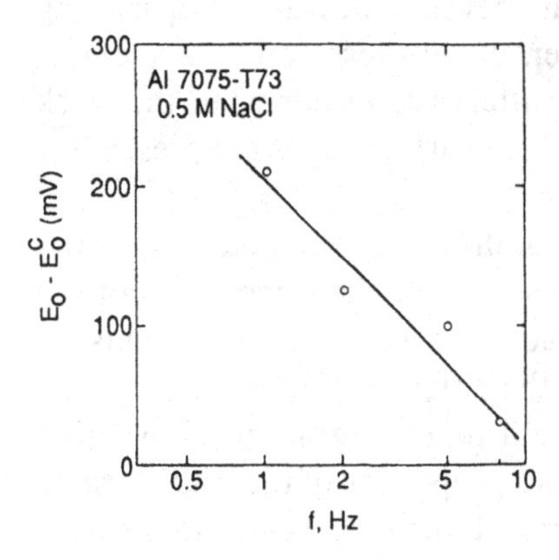

Figure 15. The difference between the open-circuit potential E_o of the bulk Al 7075 and the crack tip E_o^c as a function of the frequency of the modulation of ΔK.

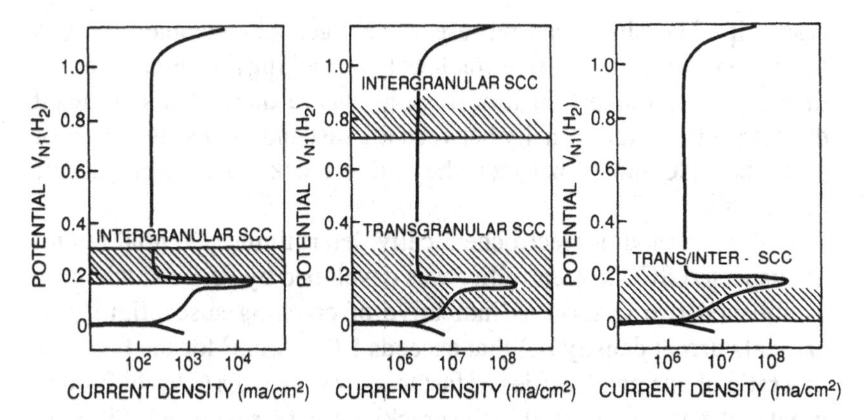

Figure 16. Potentiokinetic polarization curves and electrode potential values at which intergranular and transgranular stress corrosion cracking appears, in a 10% NaOH solution at 288°C: (a) alloy 600 (Inconel), (b) alloy 800 (Incoloy), (c) AISI 304.

This approach has been carried one step further. With the use of a slow scan rate (10 mV/min) a polarization curve demonstrating filming tendency is generated for an alloy in a specific environment. The scan is repeated at a rapid rate of approximately 1000 mV/min, reducing the chances for film formation. When these curves are compared (Fig. 17), probable potential ranges for SCC are predicted [11]. The fast scan defines those potential ranges in which high anodic activity is likely and relates to expected film-free conditions at the

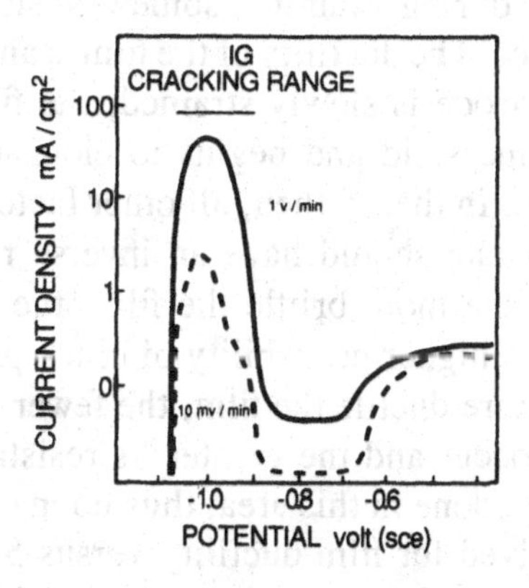

Figure 17. Polarization curves for C/Mn steel in boiling 35% NaOH.

crack tip. The slow scan represents changes accompanied by film formation relating to passive crack walls. Cracking then is expected in those active-passive transition regions where there is a significant difference in current density between the fast and slow scan curves or from the SCC model between the active crack tip and the passive crack walls.

This method of electrochemically determining potential ranges of suspected SCC susceptibility was developed by Parkins, who also established two criteria as indicative of cracking susceptibility. A critical current density (i_c) that exceeds 1.0 mA/cm^2 for the fast scan or a ratio of current densities of fast to slow scan rates (i_f/i_s) of 1000 or greater at a given potential in the cracking range is required. Either of these criteria are sufficient to establish possible SCC behavior. Although Parkins has demonstrated the validity of this approach to carbonate/bicarbonate, hydroxide, and phosphate systems, there are discrepancies in its ability to predict SCC in other systems, such as alkaline and acidic sulfide solutions.

Of course, the electrochemical approach does not address mechanical factors in SCC, but only the ranges of electrochemical potential that provide suitable conditions for SCC. Thus, a common method to confirm SCC susceptibility determined by electrochemical means is *slow strain rate testing* (SSRT) or *constant extension rate testing* (CERT), which are essentially equivalent.

The advantage of SSRT for SCC evaluation is the tendency to form films in an aggressive environment on the straining electrode, which then rupture during straining, somewhat simulating the behavior of SCC in service. The ductility of the film is an integral part of the SSRT. As the electrode is slowly strained, the film behaves as any other polycrystalline solid and begins to elongate, then thins, and ultimately ruptures. In theory then, all other factors being equal, the crack propagation rate should have an inverse relationship to film ductility. That is, the more brittle the film, the more frequently it ruptures, creating a higher probability of crack propagation into the metal, while the more ductile the film, the fewer film rupture (crack initiation) events occur and the greater its resistance to SCC. Very little work has been done in this area; thus no mathematical relationship has been derived for film ductility versus SCC propagation or resistance. For films on ferrous and aluminum alloys, fracture strains in the range of 10^{-2} have been observed. However, these represent quite brittle films, whereas other films such as ZrO_2 and Ta_2O_5, have substantial ductility. In fact, thin films of Ta_2O_5 have demonstrated up to 50% elongation without fracture. Figure 18 shows that film ductility is a strong function of thickness [13]. Current density was used in this case to represent the current necessary to repassivate areas of ruptured film. Therefore, the more film rupture sites there are, the greater is the current density required to repassivate. There appears to be some optimum film thickness that produces the best film ductility and, thus, resistance to SCC. Thick films may already contain sufficient strain for spalling to occur, thereby reducing their overall ductility.

If the dissolution rate of an alloy at a region of film rupture is assumed to be the rate-controlling process in SCC, then a maximum crack growth rate can be calculated based on electrochemical principles.

Measuring crack propagation rates with either SSRT or fracture-mechanics-type specimens and comparing these data to the dissolution current densities of film-free electrodes in the same environment gives an approximation of the *maximum* rate of crack propagation under anodic control (Fig. 19) [14]. From Faraday's second law an

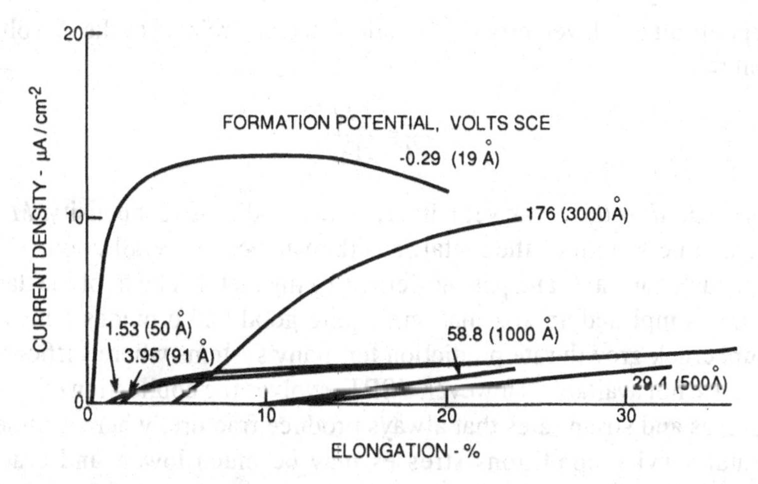

Figure 18. Current density, extension curves for zone-refined tantalum wire covered with oxide films of different thickness. Deformation potential 0.29 V (SCE), 1% ammonium borate solution.

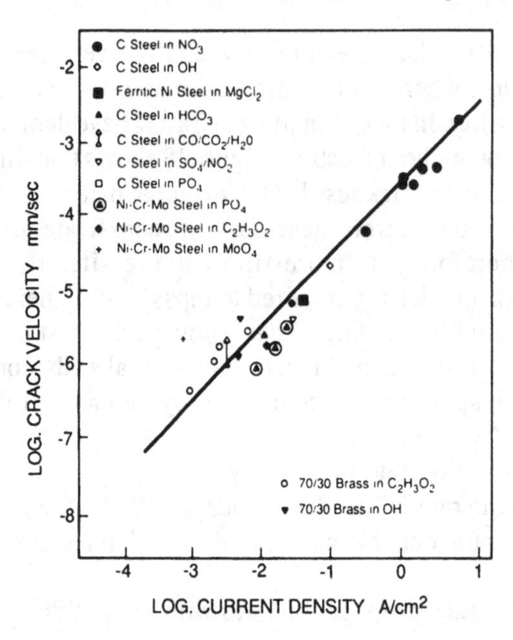

Figure 19. Measured stress corrosion crack velocities and current densities passed at a relatively bare surface for different metals in a variety of solutions.

upper limit crack velocity can be calculated and related to the dissolution rate:

$$\frac{da}{dt} = \frac{i_a M}{z F \rho}$$

where da/dt is the crack velocity, i_a is the anodic current density, M is the atomic weight of the metal, z is the valence of metallic ion, F is Faraday's constant, and ρ is the density of the metal. The fit of the data to this simplified approximation is quite good and provides a maximum crack growth rate prediction for many systems where sufficient data are not available. However, SSRT involves the application of high stresses and strain rates that always produce fracture, whereas under actual service conditions stresses may be much lower, and crack growth rate often diminishes after initial propagation at higher velocities. Thus SSRT offers a means to evaluate film-covered electrodes under severe straining conditions but is not a legitimate measure of actual or expected service conditions. It does provide a means of ranking alloys in a specific environment or to evaluate changes in the environment and the effect of these changes on susceptibility of an alloy to SCC.

Therefore, the study of SCC must carefully balance strain rate, especially crack tip strain rate, and the rate of electrochemical reactions to effectively model true susceptibility of materials to SCC. Figure 20 illustrates a threshold crack tip strain rate necessary to propagate SCC in mild steel analogous to a threshold stress intensity [1]. There appears to be a minimum crack tip strain rate necessary for SCC to occur, yet if the strain rate is too high purely mechanical failure results before electrochemical reactions can participate. A balance between breakdown of the passive film, repassivation rate, and crack tip strain rate must be achieved before SCC can occur. This balance is presented schematically in Fig. 21 [15]. As shown in Fig. 19, the fastest crack growth rate is achieved when the crack tip is continuously bare (upper curve). Conversely, when a stable film develops passivation (i_p), the crack growth rate is a minimum (often neglible). In between these extremes, crack extension occurs by repeated film rupture and repassivation. The rate of crack growth depends on the relative rate of these two mechanisms. In Fig. 21, this repetition of film rupture, crack growth, and repassivation translates to an average crack growth rate (dashed line).

A concentrated effort toward evaluating specific film properties and relating these factors to SCC has not been made. Rather, most

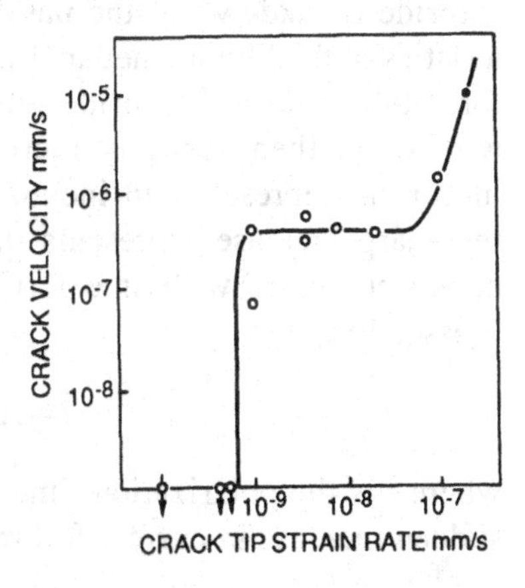

Figure 20. Intergranular crack velocities for various applied crack tip strain rates in C/Mn steel immersed in 1 N $Na_2CO_3 + 1$ N $NaHCO_3$ at -650 mV (SCE) and 75°C.

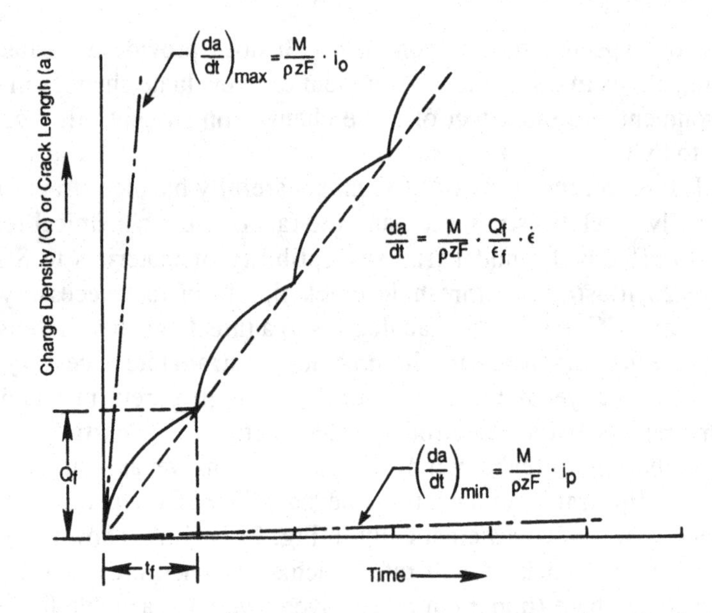

Figure 21. Charge density (Q) or crack length (a) as a function of time during stress corrosion cracking. The dashed line represents the average crack growth rate. The subscript f is for failure or fracture of the film.

investigators consider only the presence or absence of a film and do not measure or incorporate its more basic properties in their evaluations. Limited work that addresses the role of specific film properties on SCC has been performed.

For instance, the initiation of SCC in 304 stainless steel by chloride breakdown of the passive film has been related to characteristics of the film formed in different solutions. During passive film formation in neutral chloride solutions, the passive current decreases with time, then abruptly increases (Fig. 22) [16]. The decrease in current represents an incubation period for pit initiation, while the sharp increase represents the onset of pitting. The decreasing anodic current (i) with time portion of the curves in Fig. 22 can be expressed by

$$i = At^{-m} + B$$

where t is the polarization time and A, B, and m are constants. The sudden increase in current follows the relationship

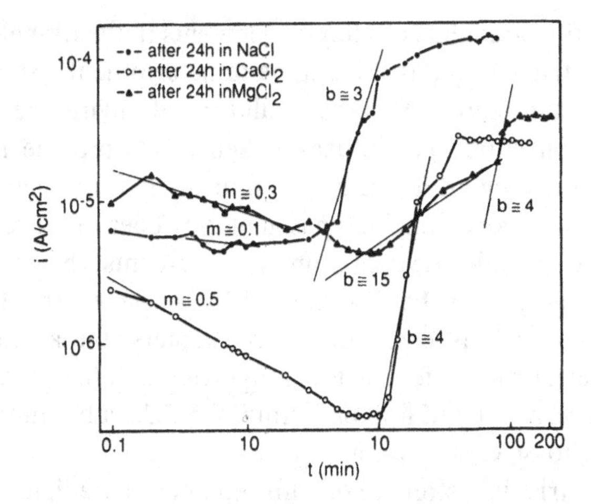

Figure 22. Plots of passive currents in moderately corroding solutions of 0.01 M (pH = 7) on 304L steel exposed previously for 24 hr in 5 N boiling NaCl, $CaCl_2$, and $MgCl_2$ solutions tested electrode potential +750 mV (SCE).

$$i = kt^b$$

where k and b are constants.

These equations are valid for neutral pH chloride solutions that are oxygen-free. The $m = 0.5$ dependence for $CaCl_2$ is indicative of ionic diffusion-controlled transport through the passive film, while $m = 0.1$ for films formed in NaCl has a predominant drift component that depends on the electric field of the film. The $m = 0.3$ exponent suggests film conductance related to both transport mechanisms. Neither the diffusion-controlled or drift-controlled species have been identified in this work. The pit induction time from Fig. 22 shows that films formed in NaCl pit faster than either of those formed in $CaCl_3$ or $MgCl_2$. Likewise, the time to failure by SCC for 304 in these solutions follows the same order, with NaCl producing the most rapid failure. Thus, the properties of the film are of considerable importance in determining SCC resistance, or at least the rate of SCC initiation. Differences in transport mechanisms through the film may have a substantal impact on the susceptibility to SCC, yet essentially no work has been done to investigate these dependencies.

Auger analyses of these films have presented a complex picture of concentration gradients of alloying elements in the film along with cations from the solution that combine to determine the structure or lack thereof (amorphous) for a particular metal-environment combination. Yet concentration profiles in themselves provide relatively little information regarding the role of films in SCC or other forms of corrosion attack. For example, it has generally been observed that for stainless steels in chloride containing environments, the chloride ion is not found deep into the film beyond the passive film/electrolyte interface. Thus, as discussed in earlier chapters, the actual role of chlorides may be their effect on film properties and, hence, the subsequent effect on an altered film on pitting or SCC, rather than directly contributing to SCC of an alloy.

The electric field strength of a film (potential gradient) may also play a part in SCC. In 5 *N* H_2SO_4 with NaCl, a high susceptibility to SCC of AISI304 stainless steel is observed, accompanied by a film of low electric field strength (0.17 mV/Å). In 0.1 *N* HCl, the electric field is quite strong (0.60 mV/Å) and SCC does not occur. For solutions of 1 *N* and 3 *N* HCl, intermediate potential gradients are observed (0.20 and 0.36 mV/Å, respectively), and SCC susceptibility of 304 is accordingly intermediate between severe and no detectable cracking. Similarly, the role of molybdenum in rapid repassivation of film-ruptured 316L stainless steel appears to be related to an increasing electric field strength. Yet Mo is frequently not found in the film itself.

For all SCC, the contribution both negatively and positively of films has not been fully evaluated. Films are most often represented in a general context as either a barrier or a permeable membrane that neither contributes to nor effects the susceptibility to SCC, except possibly as a crack initiation site. However, limited data to date suggest that film composition, structure, and electrical properties play a very real and important part in determining the SCC resistance of alloys. Moreover, the concept of specific ions causing SCC of specific alloy systems may actually be attributed to the specific ion/film interaction behavior instead. In fact, recent work on the stress corrosion cracking of ceramics suggests that the strained bonds of the oxide at a crack tip are susceptible to fracture from water molecules or other molecules of similar electronic size and structure. Therefore, data on SCC of ceramics may be applicable to oxide-covered metals.

Promotion and Inhibition of Hydrogen Damage by Films

A great deal of research over the last century has concentrated on the various forms of hydrogen damage in metals and alloys in order to understand and eventually prevent this catastrophic form of environmentally induced failure. Although significant strides in the phenomonological understanding of HD have been made, no single model currently exists that explains the varied behavior of all alloy systems in hydrogen-bearing environments.

Much of the understanding to date on HD is related to fracture initiation and propagation, which unlike SCC is predominantly subsurface or at some critical distance beyond the crack tip and associated with regions of high triaxial stress. Since hydrogen cracking is related to the arrival rate and concentration of hydrogen at the crack tip surface, films will alter the ease of hydrogen ingress in metals.

Yet with the extraordinary effort in HD research, little work has been directed to the role of surface films in both promoting and inhibiting hydrogen entry into metals. The processes of hydrogen adsorption, evolution, and absorption are well known in electrochemistry and represent the principle reactions that create the necessary conditions for HD.

Figure 23 illustrates the steps for hydrogen adsorption, absorption, and evolution [17]. In the first step, hydrated protons are transported to the double layer, at which time they are desolvated (step 2) and leave the proton adsorbed at the metal surface (step 3). Due to the electrical charge at the interface, protons are electronated to form hydrogen atoms (reduction) (step 4). After this stage, either or both of two steps may occur. The *hydrogen evolution reaction* (HER) or the *hydrogen absorption reaction* (HAR) will occur. Hydrogen recombination to form molecular hydrogen can take place by one of two mechanisms: combination of two adjacent adsorbed hydrogen atoms, or combination of a proton and adatom in the vicinity of sufficient electric charge (step 5a). Concurrent with step 5a the HAR may occur where hydrogen atoms are absorbed into the metal, giving up the one electron to the valence band of the metal, returning to its proton species in the alloy (step 5b), then diffusing to traps. The HER further

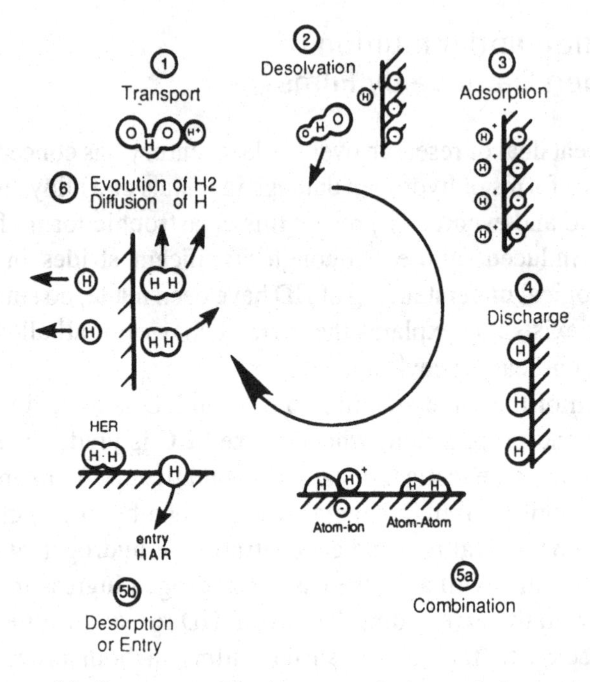

Figure 23. Processes of hydrogen evolution and absorption.

continues with a desorption step of the hydrogen molecule and transport away from the double-layer region (step 6).

While all of these steps are important to the overall corrosion rate and, under different environments (pH, temperature, velocity, etc.), have variable energetics, the absorption step appears most frequently to be the rate-determining step for many alloy systems that are susceptible to HD. However, it would be expected that the presence of surface films could sufficiently affect all of these steps to some degree, especially steps 3, 4, and 5b, and that the rate-determining step could be shifted to favor either the HAR or the HER.

The rate of absorption of hydrogen atoms on bare metal is a function of concentration times an activation energy [17]:

$$r = v\theta N_s \exp\left(\frac{-\Delta H}{RT}\right)$$

where r is the rate of entry, v is the frequency term, θ is the fraction of surface sites covered, N_s is the number of possible surface sites for absorption, ΔH is the activation energy for absorption, R is the gas constant, and T is the absolute temperature.

It is generally considered that surface coverage (fraction of sites covered) is enhanced for clean metal surfaces, with the exception of surfaces contaminated with hydrogen recombination poisons. Recombination poisons such as S, As, Sn, and Sb increase hydrogen absorption even though the surface coverage of hydrogen is reduced. However, it is expected that the activation energy for absorption could be significantly different in the presence of films as well as the frequency term, so these factors may actually increase or decrease the overall rate of absorption, compared to a bare metal surface. For example, an oxide 2–3 molecules thick on tantalum can reduce the hydrogen absorption rate by 10^{10} over a bare tantalum surface.

For iron-chromium alloys under very negative potentials and in pH 14 water, introduction of a scratch through the preexisting oxide can induce significant hydrogen evolution at the bare metal. This evolution may be prolonged and may enhance hydrogen entry into the metal, depending on reaction kinetics. Thus, even under conditions that normally promote SCC, such as film rupture, hydrogen may be a significant contributor to the cracking mechanism.

For titanium, the presence and characteristics of a protective oxide film can greatly decrease hydrogen entry into the alloy [18] (Table 1), compared to essentially bare metal surfaces. High-temperature oxidation provides a more hydrogen-resistant oxide than

Table 1. Effect of Surface Condition of ASTM Grade 2 Titanium on Hydrogen Uptake from Cathodic Charging

Surface condition	Average hydrogen pickup (ppm)
Pickled	164
Anodized	140
Thermally oxidized (677°C) (1 min)	94
Thermally oxidized (677°C) (5 min)	92
Thermally oxidized (760°C) (1 min)	82
Thermally oxidized (760°C) (5 min)	42

that achieved by anodizing, thereby reducing hydrogen uptake in the metal. Through hydrogen permeation techniques, the variability in hydrogen permeability of a membrane and the surface effects may be studied. Figure 24 shows that the more acidic the environment is, the greater is the hydrogen flux J [19]. The maximum in hydrogen flux is a concentration effect, since decreasing pH simply supplies a higher concentration of hydrogen. Both the lag time (time for measurable hydrogen flux) and J_{max} can be seen to depend on pH and are representative of hydrogen trapping behavior in the alloy [19] (Fig. 25). After the traps are filled, a steady-state permeation is achieved which depends on surface films. Thus, the steady-state permeation current should reflect the role of surface films in inhibiting or promoting hydrogen entry. However, this is difficult to unequivocally establish in aqueous solutions since, first, the solubility of a film is pH-dependent, and, second, the very nature of the film itself is pH-dependent. That is, the precipitation or nucleation of the film, its growth, and its properties are all pH-dependent. This complicates the study of hydrogen transport behavior of corrosion films in aqueous solutions at different pH.

Oxide films most often reduce hydrogen permeation; however, under some conditions they may also increase hydrogen permeation in metals. The high reactivity of bare metal surfaces often aids the hydrogen recombination reaction, thereby allowing a greater portion of hydrogen to evolve as gaseous hydrogen, reducing hydrogen absorption, compared to an oxide-covered metal. In the presence of a stable oxide, permeation of hydrogen can increase even though the

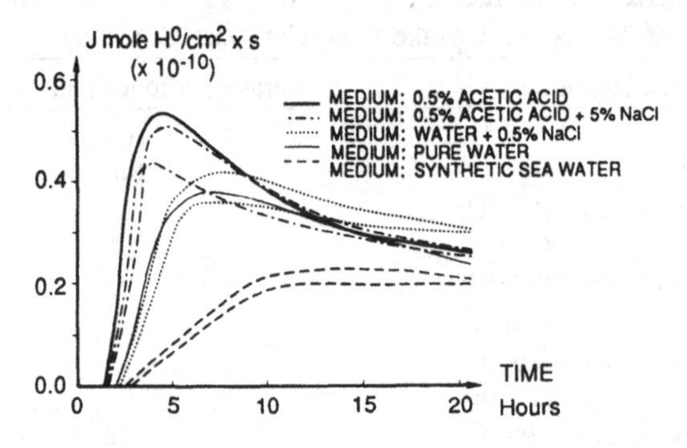

Figure 24. Effect of the test medium on the permeation curves.

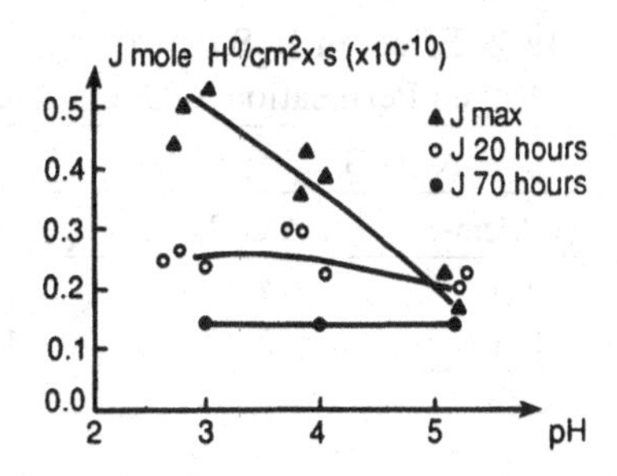

Figure 25. Evolution of permeation flux with pH.

anodic and cathodic reactions are reduced [20] (Fig. 26). The tempered surface of steel (thermal oxide) shows an increased hydrogen permeation compared to the surface that was subjected to anodic dissolution. Even though the anodic dissolution increased both the anodic and cathodic reaction rates, apparently the hydrogen recombination reaction was more efficient for the bare surface than for oxides on the same metal.

The type of hydrogenated species is important to permeation rate and critical current density (proton discharge), as illustrated in Table 2 [20]. In the presence of a preformed oxide, H_2S significantly accelerates hydrogen permeation through the oxide, compared to cathodic charging. Therefore, films can either enhance or reduce hydrogen permeation, thereby affecting the overall hydrogen embrittlement of the base metal itself.

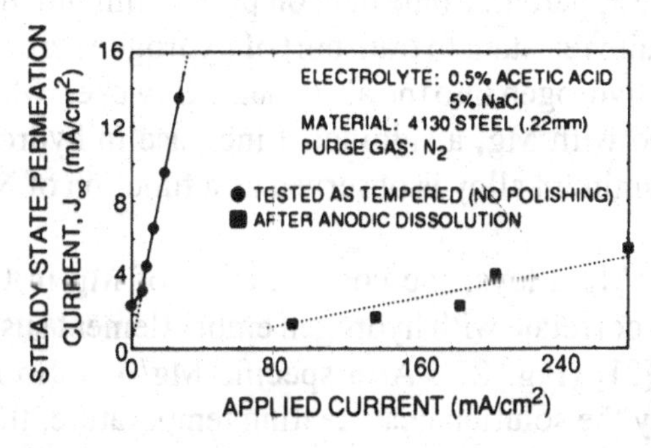

Figure 26. Steady-state hydrogen permeation after different surface treatments.

Table 2. Effect of H_2S on Corrosion and Hydrogen Permeation at Open Circuit

	No H_2S		H_2S	
Surface	i_c ($\mu A/cm^2$)	J ($\mu A/cm^2$)	i_c ($\mu A/cm^2$)	J ($\mu A/cm^2$)
Oxided	2.0	1.7	190	145
Oxide removed	8.5	1.2	180	120

Limited work has been performed to study the behavior of hydrogen within, and its transport across, a corrosion film. If protons are the predominant hydrogen species, then migration is governed by the potential gradient of the corrosion film (field strength) and diffusion is influenced by the concentration gradient. Also, the ion selectivity of the film may play a role in H^+ selectivity. As the electric field strength increases, the velocity of hydrogen ion transfer across the film increases. The diffusion mechanism of protons has been suggested to occur by hopping between adjacent O^{2-} ions, forming transient hydroxyl ions. However, if the film is an *n*-type semiconductor, protonation may take place at the oxide-environment interface, creating hydrogen atoms that would simply diffuse through the film as a result of the concentration gradient with no ionic interaction or contribution from the potential gradient. Recently, it has been found that hydrogen in the passivating film on Ni significantly reduces the resistance to pitting attack in chloride solutions.

As the understanding of SCC in aluminum alloys improves, the role of hydrogen embrittlement in the environmental cracking of Al becomes more apparent. Some data on pure aluminum metal indicate the oxide is quite resistant to transport of hydrogen across it, reducing the ingress of hydrogen into the aluminum. However, when the aluminum is alloyed with Mg, a significant increase in hydrogen permeation of the aluminum alloy is observed as a function of Mg content of the film.

In Al/Zn/Mg alloys, the concentration of Mg in the oxide has been found to correlate with hydrogen embrittlement susceptibility of these alloys [21] (Fig. 27). At a specific Mg/Al ratio in the oxide, determined by the solution heat-treating temperature, the ductility of the alloy in air saturated with water vapor reaches a minimum. This behavior is assumed to be the result of magnesium hydride formation,

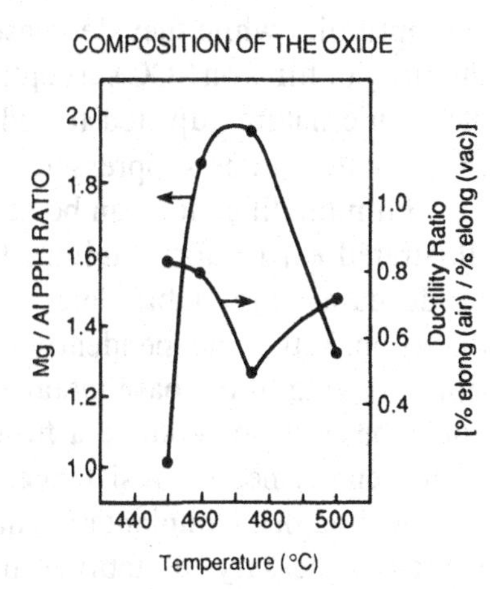

Figure 27. Ductility loss from hydrogen as a function of Mg concentration in the oxide for aluminum magnesium alloy.

which reduces hydrogen recombination at the oxide/solution interface and enhances the hydrogen content of the film. Hydrogen is generated at the oxide/solution interface by outward diffusion of Al^{3+} and reaction with water according to the reaction

$$2Al^{3+} + 3H_2O \rightarrow Al_2O_3 + 6H^+$$

Of course, the creation of MgH in the film does not in itself generate hydrogen damage of the substrate alloy. However, hydrides are quite brittle, and the creation of hydrides in the film could establish necessary conditions for brittle fracture from the oxide that correspond to the SCC mechanism of oxide-induced fracture of ductile alloys described earlier.

The Use of Films to Reduce SCC and HD

Although the technology and instrumentation are available to investigate the electronic and ionic contributions of corrosion films to SCC and HE, little has been done in this area. Detailed examination of corrosion films and perturbations from alloys and environmental changes should provide information on the generation of films with desirable properties to resist either SCC or HD or both.

In a general sense, the presence of corrosion films increase SCC susceptibility while they decrease HE susceptibility. However, if the ductility of films on SCC susceptible materials were increased and the electronic nature adjusted to reduce proton transport, then HE and SCC could both be suppressed.

Film modification can be accomplished by several methods. As illustrated earlier for titanium, thermal oxidation versus anodizing produces a more stable defect-free structure that resists hydrogen diffusion. Either independently or in conjunction with thermal oxidation, alloying of the base metal either in bulk or by ion implantation could be used to produce a film of greater ductility and increased hydrogen permeation resistance.

Although ion implantation has potential in this regard, additional work is necessary to stabilize films that contain the beneficial implanted ion from further degradation in the aggressive environment. For example, during anodic dissolution, platinum, which retards hydrogen permeation through metals, may be readily eliminated from the alloy in a very short time by corrosion [22] (Fig. 28). Once Pt ions have been depleted from the surface layers, the alloy returns to normal hydrogen permeation behavior [22] (Fig. 29). When a stable film containing Pt can be maintained, and thus the implanted region conserved, hydrogen permeation is significantly reduced for long times [22] (Fig. 30).

Thin-film technology has evolved rapidly in the last five years

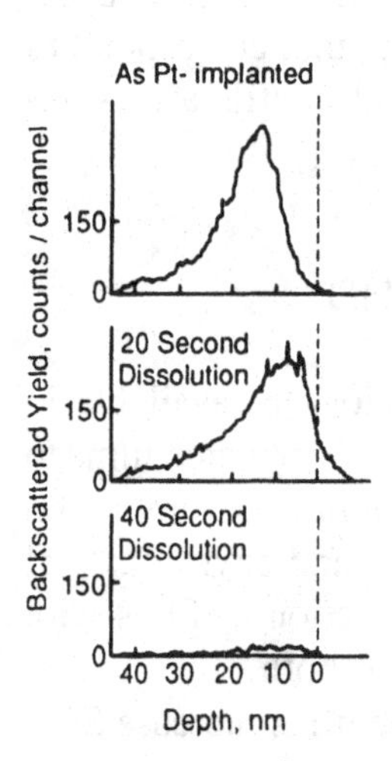

Figure 28. Rutherford backscattering profiles of Pt-implanted Fe, showing the surface concentration of Pt goes through a maximum with increasing time of immersion in 1 M H_2SO_4.

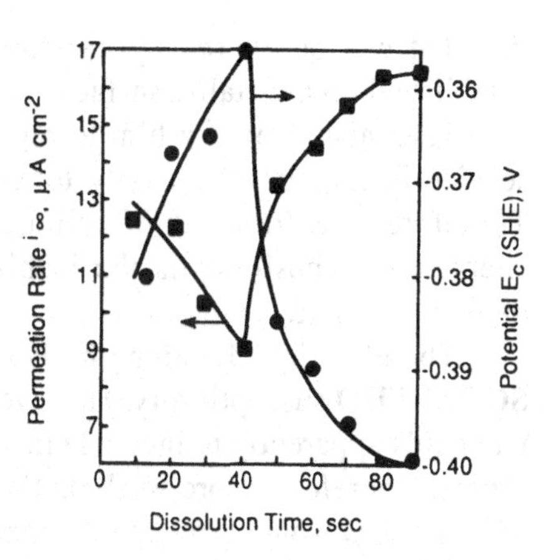

Figure 29. Steady-state permeation rate (■) and charging potential (●) for a Pt-implanted Fe membrane as a function of additional immersion periods in 0.1 *N* H_2SO_4 beyond the initial 20-sec immersion in 1 *M* H_2SO_4; $i + 0.6$ mA cm^{-2}.

and may provide the basis for developing stress-corrosion-resistant films that can be grown on alloys with the properties to resist SCC and HD. In order to produce these films, several factors must be considered. Proper attention must be paid to the defect structure of the film, its electronic conductivity and ionic conductivity under the conditions of exposure, the solubility of the film in its intended environment, and its interaction with aqueous and gaseous components of that environment. Furthermore, the diffusion of hydrogen across these films, their change in properties with hydrogen solubility in the films, and the ability to reduce hydrogen transport and solubility are all proper-

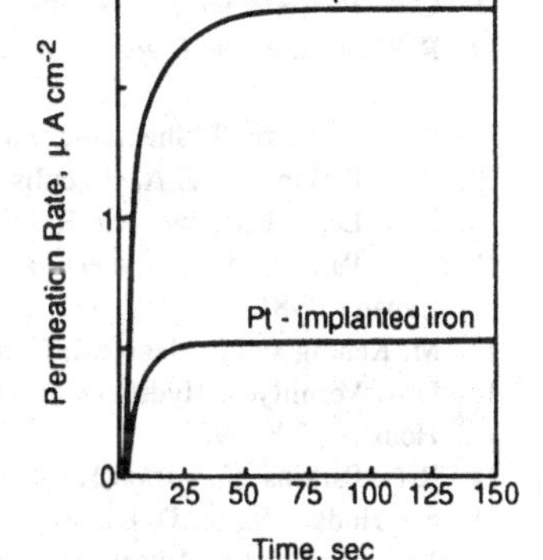

Figure 30. Hydrogen permeation transients for medium condition in Fig. 28; $i = 0.1$ mA cm^{-2} in 0.1 *N* NaOH.

ties that can be altered by careful selection of the corrosion film established on a metallic surface.

The capability of a film to repair itself after being damaged in service is important not only to SCC resistance but also for HD. Therefore, one drawback of a film grown on a substrate of different chemical compositions is the inability in situ to adequately repair damage to itself.

The electric field strength also may provide a means to reduce SCC and HD susceptibility. However, from limited data, weak electric fields apparently reduce HD in some alloys but increase SCC for others. Therefore, more work on the relationship of field strength to SCC and HD would provide information on optimum field strength to resist either of these cracking mechanisms.

The ductility of films enhances both SCC and HD resistance by reducing the entrance of brittle fractures from the film to the matrix and the number of potential sites for bare crack tips that promote SCC. If hydrogen absorption is also enhanced by the bare surfaces in the presence of corrosion films, then ductile films will produce fewer sites for hydrogen entry.

References

1. R.H. Jones and R.E. Ricker, Corrosion, vol. 13, p. 145, *ASM Metals Handbook,* 1987.
2. E.D. Hondros and C. Lea, *Nature* **1981,** 289.
3. E.D. Hondros, *J. Phys.* **1977,** 36 Coll 1C4-117 and D. Gupta, *Met. Trans.* **1977,** *8A,* 1431.
4. B.D. Craig, *Met. Trans. A* **1982,** *13A,* 907.
5. P. Doig and P.E.J. Flewitt, *Acta Met* **1978,** *26,* 1283.
6. G.G. Burstein and J.T. Woodward, *Met. Sci.* **1983,** *17,* l11.
7. R.W. Staehle, *The Theory of Stress Corrosion Cracking in Alloys,* p. 233, NATO, Brussels, 1971.
8. R.C. Newman, T. Shahrabi, and K. Sieradzki, *Scripta Met.* **1989, 23,** 345.
9. R.N. Parkins and Z.A. Foroulis, Corrosion/87, Paper No. 188, San Francisco, CA, 1987.
10. A.H. Le, B.F. Brown, and R.T. Foley, *Corrosion* **1980,** *36,* 673.
11. R.N. Parkins, P.W. Slattery, and B.S. Poulson, Corrosion/81, Paper No. 96, Toronto, Canada, 1981.
12. M. Kendig and F. Mansfeld, Corrosion/86, Paper No. 190, Houston, TX, 1986.
13. D.A. Vermilyea, Hydrogen Embrittlement and Stress Corrosion Cracking, p. 15, NACE, Houston, TX, 1967.
14. R.N. Parkins, *Corrosion Sci.* **1980,** *20,* 147.
15. S.J. Hudak, Jr., Ph.D. Dissertation, Lehigh University, 1988.
16. J. Gluszek and K. Nitsch, *Corrosion Sci.* **1982,** *22,* 1067.

17. R.D. McCright, Stress Corrosion Cracking and Hydrogen Embrittlement of Iron Base Alloys, p. 306, NACE, 1973.
18. R.W. Schutz and L.C. Covington, *Corrosion* **1981,** *37,* 585.
19. D. Petelot, M.F. Galis, and A. Sulmont, Corrosion/86, Paper No. 165, NACE, Houston, TX, 1986.
20. B.J. Berkowitz and H.H. Horowitz, *JES* **1982,** *129,* 468.
21. R.K. Viswanadham, T.S. Sun, and J.A.S. Green, *Corrosion* **1980,** *36,* 275.
22. H.W. Pickering and M. Zamanzadeh, Hydrogen Effects in Metals, TMS-AIME, Jackson Hole, WY, 1981,

7

Role of Films in Erosion-Corrosion

Introduction

Erosion-corrosion is often defined as the accelerated attack of a metal due to the relative motion between a corrosive fluid and the metal surface. The corrosive fluid need not be limited to aqueous solution but may include gases, organic fluids, molten salts, and liquid metals. Likewise, combinations of these fluids can induce other forms of erosion. For example, gases in liquid can create cavitation damage, liquid in gases can cause liquid drop impingement, and solids in gases or liquids can cause abrasion.

For environments where the mechanical aspects of deterioration are more profound than the chemical (corrosion) aspects, many investigators refer to this form of degradation as erosion-corrosion. Conversely, if the corrosion contribution is greater than the mechanical factor, the term *corrosion-erosion* is applied. However, in actual practice it is quite difficult to separate the contribution from each form of degradation; therefore, erosion-corrosion has become a general term used to identify corrosion effected by flow, and *vice versa*. For convenience and consistency with the majority of the literature, the term *erosion-corrosion* will be used. In this context, erosion-corrosion will be defined as the mechanical erosion of surface films on a metal and the subsequent deterioration of the metal by the combined influence of corrosion and mechanical damage.

The role of surface films in erosion-corrosion is quite important. Many metals and alloys depend on passivity or a surface film of some type to limit corrosion in specific environments. The removal or destruction of this film can seriously affect the performance of an alloy in an otherwise benign environment. Under erosion-corrosion conditions the ability of a surface film to protect the substrate depends on many factors. Some of the important factors are the speed with which a film forms in the environment upon initial exposure of the metal, the resistance of the film to damage from the eroding medium, and the ease or speed of film repair once damage has occurred. At first glance a hard, tenacious film seems desirable; however, it will be seen that a truly erosion-resistant film must maintain a certain degree of ductility. Brittle films will crack or spall, creating local regions of film loss which are then susceptible to accelerated corrosion. Conversely, a soft ductile film will be easily eroded, offering little resistance to the mechanical impact of the eroding particles. Thus, the ideal film will be specific to the particular erosion-corrosion environment displaying greater resistance to either the mechanical or electrochemical aspect of the environment as needed (i.e., whether corrosion-erosion or erosion-corrosion dominates).

Erosion-corrosion is a very complex subject that entails the study and understanding of the hydrodynamic aspects of a system as well as the metallurgical characteristics of the alloy. In systems where pure erosion predominates, surface films have little contribution. However, for erosion-corrosion surface films are most important to the overall appreciation of the process and thus are emphasized in this chapter. Yet, other factors are significant and must be considered for any complete understanding of erosion-corrosion.

General Aspects of Erosion-Corrosion

There are several mechanisms for erosion-corrosion, depending on whether the eroding agent is gas in liquid (cavitation), liquid in gas (impingement), or solid particulate in gas or liquid (abrasion). These different forms of erosion-corrosion have many similarities and numerous differences. The mechanisms are quite complex and involve many factors other than the surface film. However, these factors are addressed elsewhere. Regardless of the specific circumstance, typical erosion-corrosion behavior is demonstrated in Fig. 1. At some critical velocity or breakaway velocity the corrosion rate abruptly accelerates due to local breakdown of the corrosion film. At the highest velocities, a higher steady-state material loss rate is achieved, which is primarily due to pure erosion of the metal with no corrosion component. This behavior is illustrated for the case of cavitation in Fig. 2 [1]. The collapse of gas bubbles damages the film, causing increased corrosion in those areas of film damage. Increasing the number of locations of film damage (curve c) greatly magnifies the corrosion rate, compared to fewer damage sites (curve b). Thus, the overall rate of degradation is a function of both mechanical damage of the film by cavitation and corrosion from oxidation of the substrate (alloy). This process can be considered a short circuit, compared to the much slower corrosion kinetics typically observed after the film is established. In many cases, fractures or cracks in the film are as detrimental to overall degradation as complete removal of the film, because accelerated attack will be stimulated at these breaks.

Figure 3 illustrates changes in electrochemical parameters (passive current density, passive potential range, etc.) with and without vibration [2]. Cavitation created by vibration increased the passiva-

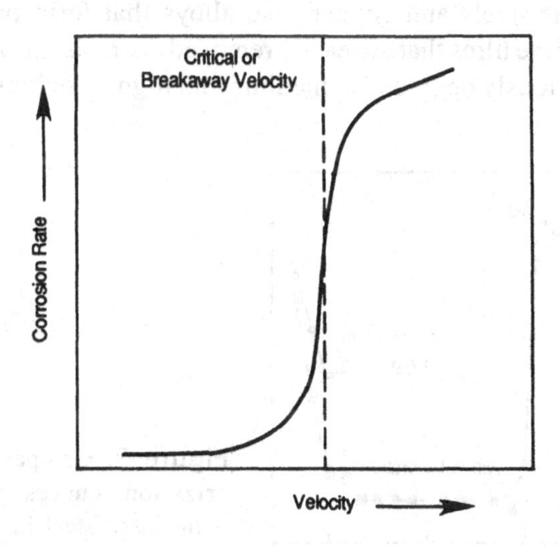

Figure 1. Ideal representation of the breakaway or critical velocity associated with erosion-corrosion.

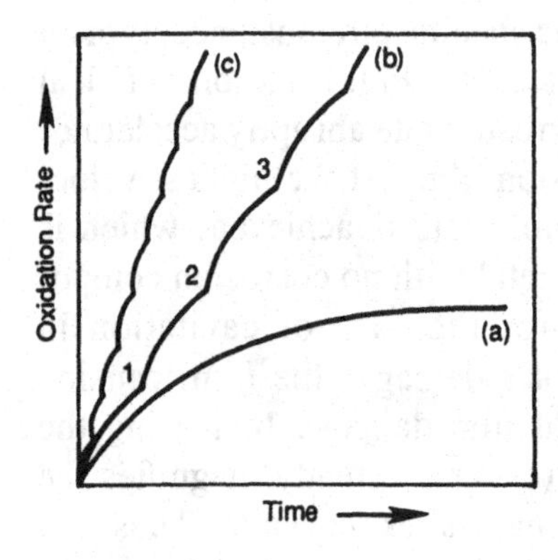

Figure 2. Cavitation damage of a corrosion film damaged at 1, 2, and 3 (b) in many places (c) and a normal film with no damage (a).

tion current about 30 times over the behavior observed for static conditions and diminished the width of the potential range over which passivity occurred.

The nature of the surface film is of paramount importance to erosion-corrosion resistance. It has been found that the erosion-corrosion resistance of alloys often fall into three categories, depending on their film forming characteristics [3]. Alloys with very tenacious protective films, such as titanium alloys and some Ni/Cr/Mo alloys, exhibit excellent erosion-corrosion resistance in seawater. Alloys such as steels and copper-base alloys that form protective or semiprotective films that are easily removed are resistant at low velocities but seriously degraded at medium and high velocities. The inter-

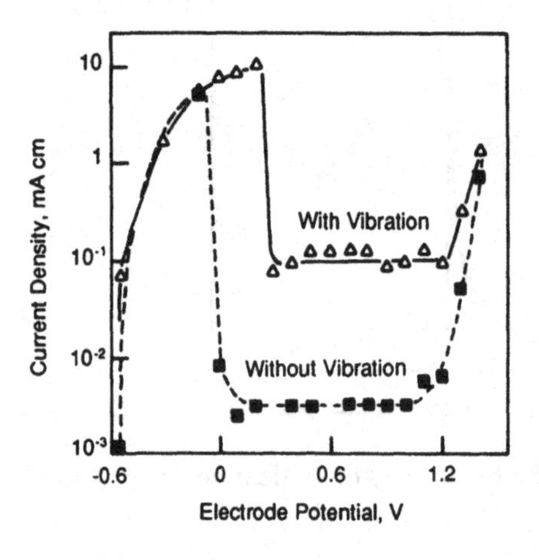

Figure 3. Comparison of polarization curves for carbon/manganese steel in a potassium/hydrogen phthalate solution with and without vibration.

mediate category of erosion-corrosion behavior is displayed by stainless steels and many nickel-base alloys. Under high and medium velocities erosion-corrosion rates are minimal; however, under static or low-flow conditions debris can settle on the passive film, creating sites for pitting or crevice attack.

The erosion-corrosion of 70/30 copper/nickel in seawater demonstrates critical velocity behavior under certain test conditions, depending on the test method (Fig. 4) [4]. For example, classical breakaway velocity is observed for the low- and medium-iron-containing alloy tested by the rotating spindle. However, other test methods do not demonstrate an abrupt change in erosion-corrosion resistance. This same figure suggests that increasing the iron content of 70/30 copper/nickel enhances erosion-corrosion resistance, including the depression of a critical velocity. For example, with the rotating spindle test at low Fe content (0.04% Fe) the alloy shows definite breakaway behavior. However, at high Fe content (5.38% Fe) the rate is insignificant. These data are not sufficient to make broad generalizations about the benefit of Fe in the erosion-corrosion resistance of 70/30 copper/nickel, but they do imply that factors other than just the mechanical aspects of erosion are at play. The small variation of iron content will not appreciably alter the mechanical properties of copper/nickel alloys, but can profoundly effect the electrochemical behavior of such an alloy. Moreover, if alloy composition effects erosion-corrosion behavior from an electrochemical standpoint, then all other factors that influence electrochemical processes will contribute to erosion-corrosion. Some of the more important factors are temperature, pH, velocity, intermittent versus continuous flow, and solution composition.

Figure 5 illustrates some of the processes that may occur for copper alloys exposed to flowing seawater [4]. Of course, this schematic can also be generalized for most metal/fluid systems under flowing conditions. This figure assumes a continuous pore-free, adherent, insoluble corrosion film. In this case, corrosion will be controlled by one or more steps, including the kinetics of transport of oxygen, CO_2, and complexing agents from the bulk solution through the turbulent boundary layer, laminar sublayer, diffusion boundary layer, and the Helmholtz double layer to the film/electrolyte interface.

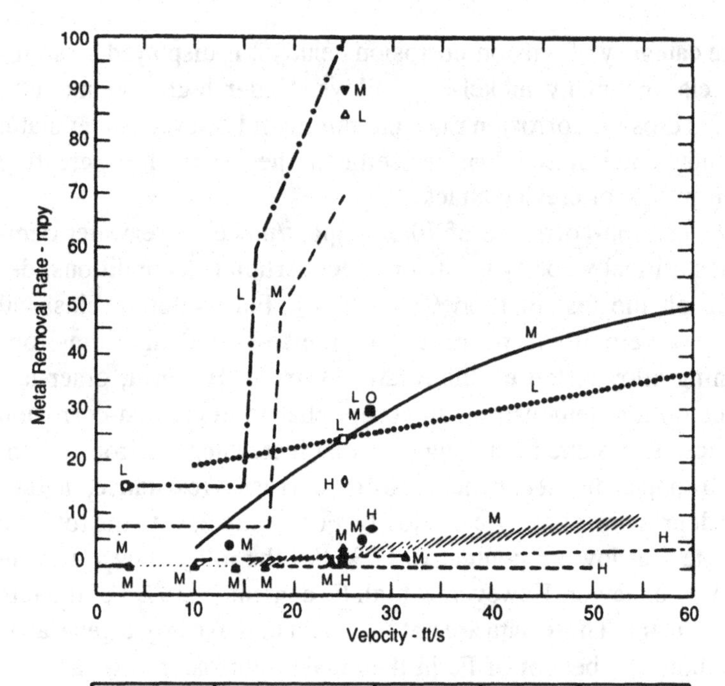

	% Fe	Test Method	Testing Time (Days)	Temperature (°C)	Reference
△	0.06	Jet Impingement	60	Ambient	6
○	0.06	Rotating Disc	60	Ambient	6
••••••	0.06	Multivelocity Jet	30	Ambient	6
□	0.06	Rotating Spindle	60	Ambient	6
—•—	0.04	Rotating Spindle	60	26	8
□	0.02	•	365	2-30	9
▲	0.4 1.0	Simulated Service (Flow Through Tubes)	208	15	3
▬	0.42	•	365	2-30	9
▼	0.6	Jet Impingement	60	Ambient	6
■	0.6	Rotating Disc	60	Ambient	6
●	0.5	Rotating Disc	60	15	10
——	0.6	Multivelocity Jet	30	Ambient	6
♦	0.6	Rotating Spindle	60	Ambient	6
— —	0.47	Rotating Spindle	60	26	8
////	0.5	•	•	Ambient	11
----	5	•	•	Ambient	11
◊	5.38	Jet Impingement	60	Ambient	6
●	5.38	Rotating Disc	60	Ambient	6
—··—	5.38	Multivelocity Jet	30	Ambient	6
■	5.38	Rotating Spindle	60	Ambient	6

Figure 4. The effect of velocity or erosion-corrosion rate on 70/30 copper/nickel alloys in seawater. L, M, and H designate low, medium, and high iron content of the alloys.

Likewise, kinetics may be controlled by diffusion of Cu or Fe cations away from the metal/film interface through the film to react at the

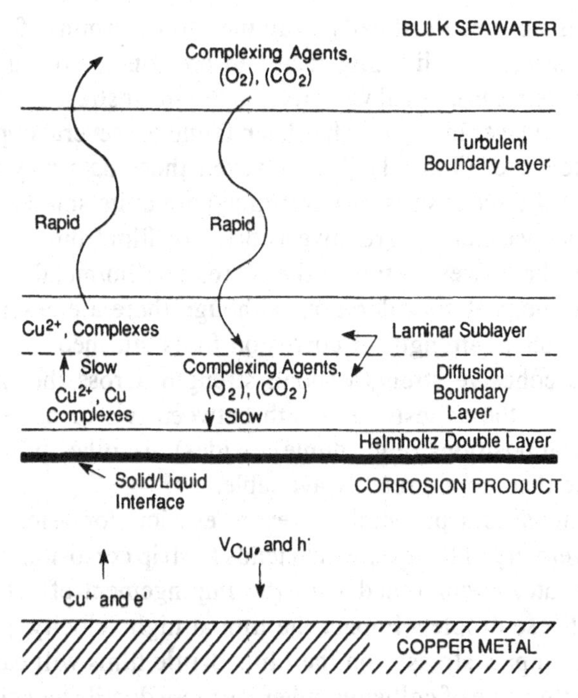

Figure 5. Representation of some of the processes that may occur at the surface of a copper alloy in solution.

Helmholtz double layer. As flow increases, the laminar sublayer and diffusion boundary layer will diminish. As this occurs, the concentration of cuprous and cupric ions decrease, thereby effectively increasing the slope of the concentration gradient and stimulating further dissolution of copper.

If the corrosion product is discontinuous, porous, somewhat soluble, or easily spalled, the transport kinetics across the film will become less important to the overall erosion-corrosion behavior than the anodic reactions at the exposed bare metal locations.

As presented in Chapter 2, when the fluid velocity increases, the diffusion layer decreases, enhancing mass transport to the Helmholtz double layer. However, at some velocity it would be expected that the shear strength of the corrosion film to the metal surface will be exceeded and the corrosion film removed.

This approach to critical erosion velocity, while intuitively satisfactory, has not been critically examined to determine if corrosion film shear stress is a limiting factor in erosion-corrosion. A good correlation between critical velocity and the shear stress created at the film/diffusion layer interphase has been found for several copper-base alloys in seawater (Table 1) [5]. However, these data may be fortuitous, since the shear stresses determined are quite small and would appear to be insufficient to remove a corrosion film from its substrate. In fact, the shear stress acting at the corrosion film/metal interface is the more important consideration, although there are essentially no data on the shear strength of corrosion films attached to metal surfaces. The cohesive strength (bond strength across the film/metal interface) and the adhesive strength between two or more films of different compositions (e.g., duplex oxides) are other properties of importance that are not readily available.

Wet steam is a particularly severe erosion-corrosion medium. The high velocity of the gas, which tends to strip corrosion films from their substrate, is combined with the impingement effect of water droplets. These droplets impact the film at high velocity, producing substantial impact stresses on the film. Brittle films will easily fracture under this type of collision, whereas more ductile tenacious films will be quite resilient. In fact, the impact of liquid droplets on corrosion films is not much different from the impact of solid particles carried by a liquid or a gas. The major difference is in the deformation of the particle itself, liquids deforming much more readily than solids. To a first approximation, these differences are not important in examining the role of the corrosion film in erosion-corrosion; therefore, the difference between abrasive erosion-corrosion (due to entrained hard particles) and liquid impingement will not be distinguished in this chapter; instead the performance of the film itself will be emphasized.

Properties of the Film and Its Relation to Erosion-Corrosion Resistance

The behavior of films in erosion-corrosion environments has been discussed from a general overview, assuming the specific properties of the film contribute essentially nothing to the resistance of the

Table 1. Critical Velocity and Shear Stress for Copper-Base Alloys in Seawater

Alloy	Critical velocity m/s (ft/s)	Temp. C	Critical shear stress N/m^2 (lb_f/ft^2)
CA 122	1.3 (4.4)	17	9.6 (0.2)
CA 687	2.2 (7.3)	12	19.2 (0.4)
CA 706	4.5 (14.7)	27	43.1 (0.9)
CA 715	4.1 (13.5)	12	47.9 (1.0)
CA 722	12.0 (39.40)	27	296.9 (6.2)

film/metal system. However, it is naive to consider that ionic and/or electronic transport across a film, the strength and ductility of the film, and the field strength of the film are not participating in the erosion-corrosion process. Moreover, some of the factors may be rate-controlling under certain circumstances.

Figure 6 illustrates the effect of potential and fluid velocity on the film formation current [6]. Increasing velocity at constant potential reflects a higher current for film formation. Moreover, the film formation efficiency decreases with increasing velocity except for the lowest velocity condition (Fig. 7). Since no film properties were measured during these tests, important factors such as film thickness and composition cannot be related to the erosion-corrosion behavior. However, under the specific experimental conditions used for these

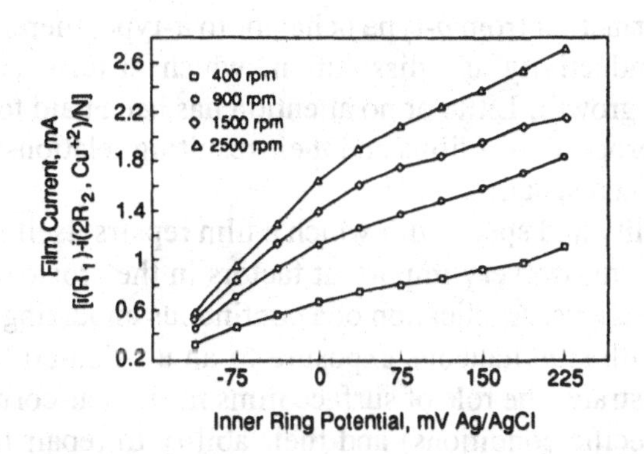

Figure 6. Effect of relative solution velocity on corrosion film current $[i(R_1) - i(2R_2, Cu^{2+})/N]$ as a function of applied potential in 1 N $NaHCO_3$ solution.

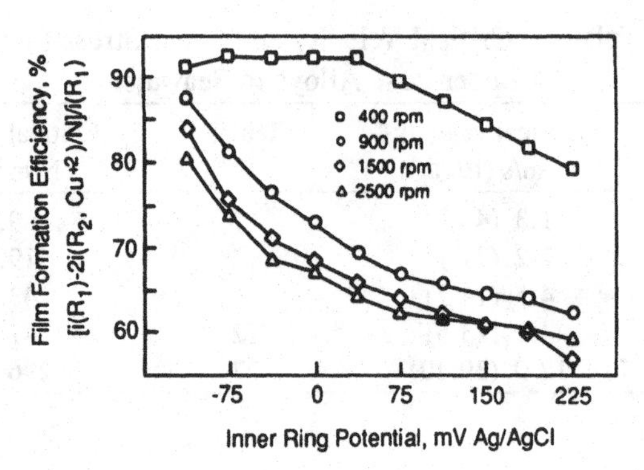

Figure 7. Similar to Fig. 6 but corrosion film formation efficiency plotted against applied potential.

data, the oxidation of cuprous ions to cupric ions occurs at about −350 mV versus Ag/AgCl. Thus, at low applied potentials the film is expected to be composed primarily of Cu_2O. As the potential increases to more positive values, the cuprous ions in the film adjacent to the film/electrolyte interface will be oxidized to cupric ions, changing the electronic properties of the film and the solubility of the film in the fluid. Since the solubility product of CuO is about 100 times larger than Cu_2O, the ability to maintain both a protective film and a thick film will be diminished with increasing potential. From an electronic standpoint Cu_2O is a *p*-type semiconductor and CuO is a *n*-type semiconductor. In general, electronic resistance of a film decreases upon transformation from *p*-type behavior to *n*-type, thereby enhancing ionic conductivity and dissolution, which in turn reduces film stability and growth. Little or no attention has been paid to the semiconductive properties of films and their resulting relationship to erosion-corrosion resistance.

The ability and speed with which a film repairs itself once it has been eroded are two very important factors in the erosion-corrosion resistance of alloys. Application of a continuous anodizing current to aluminum with simultaneous exposure to an erosion-corrosion medium demonstrates the role of surface films in erosion-corrosion (under these specific conditions) and their ability to repair themselves under erosion conditions [7]. When only the natural aluminum oxide on aluminum is present and no anodizing current is applied, the resistance of this anodic film is excellent over a wide range of fluid velocity (Fig. 8). This implies that the velocities were not sufficiently high to reach or exceed a critical velocity, and that whatever film damage occurred was readily repaired by the alloy in this environment. However, with the inclusion of abrasive particles there is an increasing rate of film destruction and metal loss with increasing velocity and an apparent critical limit somewhere between 28.2 L/hr and 52.8 L/hr.

When an anodizing current is applied and the data examined at constant time, there is obvious critical velocity behavior demonstrated by the reduction in film thickness (Fig. 9) and cell voltage (Fig. 10). Without abrasion from solid particles, a critical velocity is not achieved; however, increasing fluid velocity reduces film thickness and cell voltage, suggesting that film growth and stability are increasingly diminished. In the presence of abrasive particles, a critical velocity is reached where film repair is not rapid enough to prevent metal wastage.

In this example, there is no change of oxidation state, as was observed with copper ions. Thus, the solubility of the film is constant, as are the electronic and ionic mobilities, at least as related to specific film composition. Once the film begins to thin from erosion at constant anodizing current, the field strength across the film will increase, thereby stimulating ionic and electronic transport across the film and increasing corrosion. Differences in film thickness and composition have been observed to impart different potentials to the film [8]. In other words, erosion of the film can create highly localized potential differences as a result of variations in film thickness. This variation in potential laterally across the film will enhance the prefer-

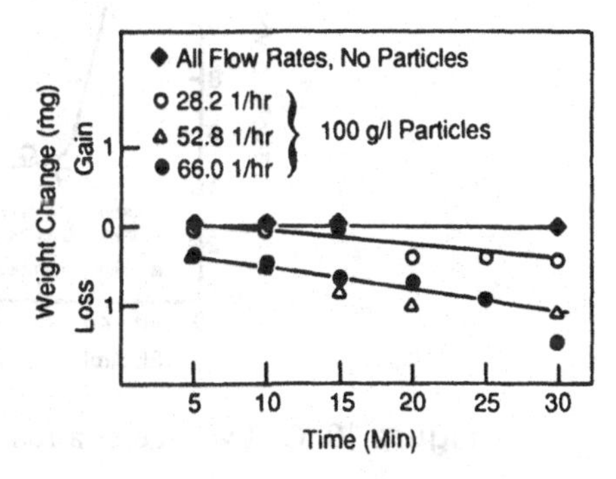

Figure 8. Weight change of aluminum as a function of exposure time to solution with and without particles in the absence of an anodizing current.

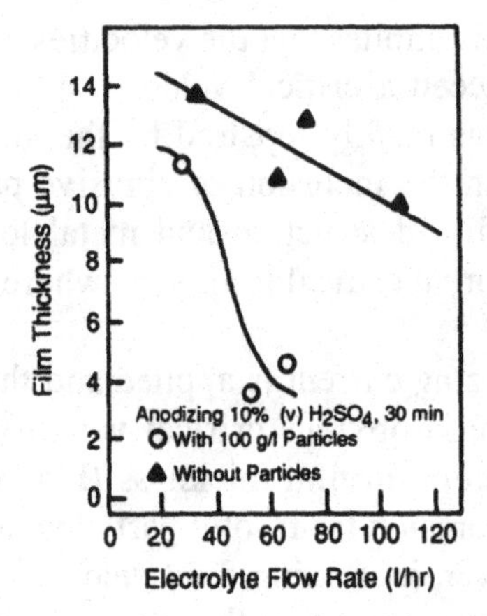

Figure 9. Corrosion film thickness as a function of solution flow rate.

ential formation of some oxides over others and promote local anodes and cathodes that may stimulate further degradation of the film.

The film composition has a significant effect on the erosion-corrosion resistance of a material. For example, Fig. 11 illustrates the decreasing erosion-corrosion rate of steels in neutral pH, deoxygenated water at 355°F (180°C) as a function of alloy content [9]. Increasing the alloy content of the metal enhances the erosion-corrosion resistance as a result of altering the corrosion film composition. Chro-

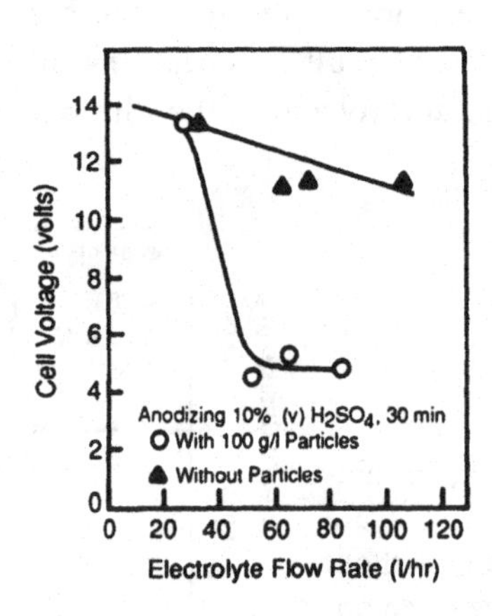

Figure 10. Cell voltage as a function of solution flow rate.

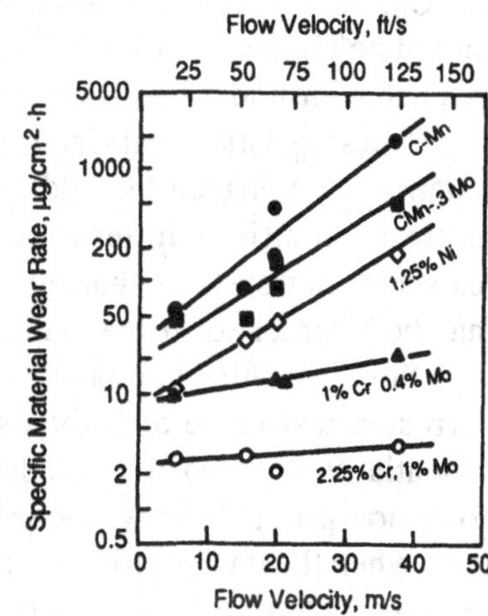

Figure 11. Effect of flow rate and alloy composition on erosion-corrosion rate of alloys in high-temperature, high-pressure water (180°C, 4 MPa).

mium is particularly effective in reduced erosion-corrosion, even at relatively low concentrations.

For gas systems containing CO_2, the corrosion product has been found to contain an increasing concentration of Cr as the Cr content of the alloy is increased up to some threshold [10] (Fig. 12).

It was determined that elevated levels of Cr in the film were present as a Cr^3—O compound and that the corrosion film contained Cr_2O_3. Therefore, Cr^{3+} ions probably enter the film and combine as a spinel in the $FeCO_3$ matrix until some solubility limit is exceeded and Cr_2O_3 precipitates. Since Cr_2O_3 is a harder, more tenacious film than

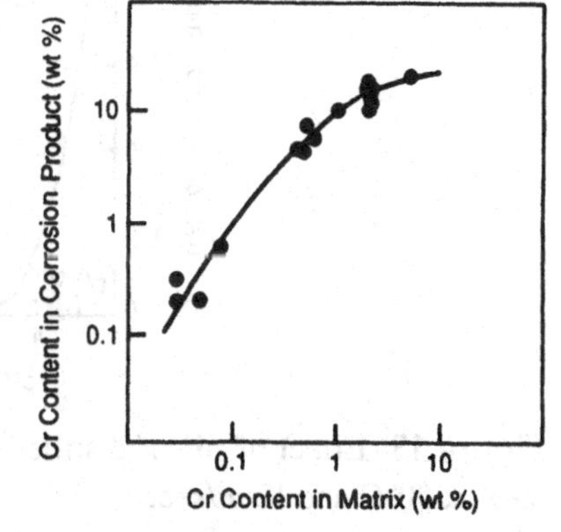

Figure 12. The relationship between base metal Cr content and corrosion film Cr content upon exposure to 0.1 MPa CO_2 at 60°C.

$FeCO_3$, the erosion-corrosion resistance of films with increasing Cr should be enhanced. Limited data indicate this is true. Examples of this behavior under high-temperature erosion are presented later.

The suggestion that Cr_2O_3 should enhance erosion-corrosion resistance is confirmed by field experience, which shows higher chromium austenitic stainless steels have greater erosion-corrosion resistance in wet steam than do low-alloy steels, even though the latter may be a harder substrate than the austenitic stainless steel.

Similarly film composition has been found to effect erosion-corrosion resistance of carbon steel in distilled water [11] (Fig. 13). Essentially no corrosion was detected at pH 6 and pH 10, where the corrosion products were composed of $Fe(OH)_2$ and $Fe(OH)_3$. However, when the composition of the corrosion film changed to Fe_3O_4 at pH 8 and below pH 6, the rate of attack was significant. The corrosion product below pH 5 was observed to readily form cracks, exposing bare metal to the fluid. Of course, at low enough pH values, the corrosion film will become increasingly soluble in the fluid until sufficient saturation necessary for precipitation at the interface cannot be maintained, and corrosion progresses unabated.

For iron or steel in oxygenated water, erosion-corrosion has been related to the mass transfer rate across the diffusion boundary. The ions transferred are either ferrous ions created at the metal/film interface and transported across the film to the film/solution interface, where they become dependent on mass transfer at this location, or

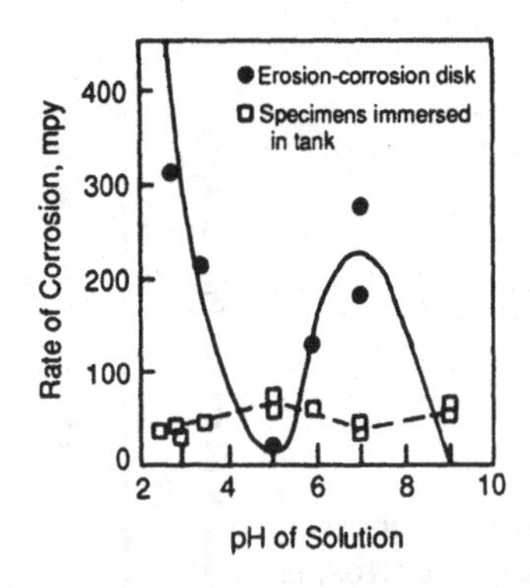

Figure 13. Effect of pH of distilled water on erosion-corrosion of carbon steel at 50°C and 12 m/sec.

dissolved oxygen in the bulk solution, which by mass transfer is concentrated at the solution/film interface.

It has been found in a series of studies that the wear rate in slurry systems follows the relationship [13–15]

$$W = Av^n$$

where W is the wear rate, v is the velocity, and A and n are constants. When n ranges from 2 to 3, metal wastage is predominantly by erosion. However, when n varies between 0.8 and 1, oxygen mass transfer controls the corrosion rate. The value of A depends on the specific system and may be calculated from consideration of the Sherwood, Reynolds, and Schmidt numbers. Figure 14 shows the erosion-corrosion rate, assuming that ferrous ions were oxidized chemically by oxygen to form $Fe(OH)_2$ within the diffusion boundary layer (curve 6) or that ferrous ions were oxidized in the bulk solution outside the diffusion layer (curve 5). It would appear that most of the reaction with ferrous ion occurs within the diffusion boundary layer. However, this analysis does not explain whether the reaction of ferrous ions is occurring at the film/solution interface or at eroded locations where the ferrous ions could be entering the solution directly at the metal/solution interface. Furthermore, since the diffusion of ferrous ions through the film should be the rate-limiting process compared to oxygen mass transport across the boundary layer, there should be a considerable influence of the film composition and proper-

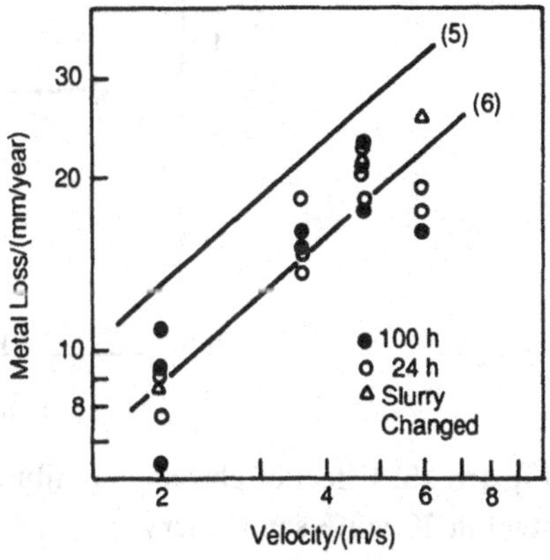

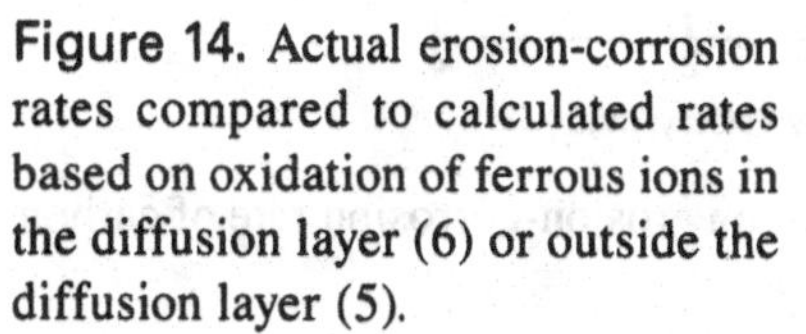

Figure 14. Actual erosion-corrosion rates compared to calculated rates based on oxidation of ferrous ions in the diffusion layer (6) or outside the diffusion layer (5).

ties on erosion-corrosion rate that are not reflected in the fluid mechanics approach.

If the stability of the oxide can be increased by application of some chemical corrosion inhibitor, then the corrosion component should be reduced simply because all the reactions would be relegated to transport across the film, which will be slower than for reaction at the bare metal/solution interface. Figure 15 is a good example of the beneficial effect of inducing the formation of an erosion-resistant film by addition of an inhibitor [15]. With no inhibition the erosion-corrosion of carbon steel in a slurry of sand in aerated tap water displays breakaway behavior, signaling loss of the corrosion film. When potassium chromate is added to the 8.5-pH water, a substantial reduction in erosion-corrosion is observed, and breakaway is not reached within the velocity regime examined.

Chromates are anodic inhibitors, but their effectiveness is highly pH-dependent. For example, for pH values less than 3 the following reaction occurs:

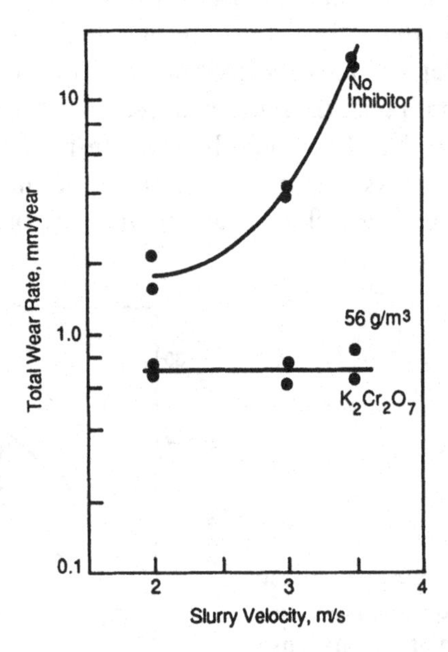

Figure 15. Effect of chromate inhibitor on erosion-corrosion rate of carbon steel in 20 vol% sand slurry.

$$Cr_2O_7^{2-} + 7H^+ + 6e \rightarrow 2Cr^{3+} + 7OH^- \quad (1)$$

The reduction of chromate predominates, which may actually stimulate corrosion. However, at higher pH, chromates may reduce to chromium oxides or hydroxides:

$$Cr_2O_7^{2-} + 4H^+ + 6e \rightarrow Cr_2O_3 + 4OH^- \quad (2)$$

The formation of Cr_2O_3 produces a protective film that resists erosion-corrosion, as mentioned earlier for austenitic stainless steels. Thus, inhibitors may aid in erosion-corrosion resistance not only by inhibiting the corrosion reaction but also by promoting the formation of protective films. However, the specific film characteristics are only important in erosion-corrosion behavior as long as transport across the film is the only, or primary, means for ionic movement into the solution. Once the film is damaged, mass transport of oxygen to the bare metal/solution interface becomes controlling and is more a function of fluid dynamics than of film characteristics.

Since mass transport of oxygen can have a large effect on erosion-corrosion rate, the elimination or reduction of oxygen content in the fluid should reduce the erosion-corrosion rate. In fact, this is what is observed for low and ambient temperature conditions. However, the reverse is true at higher temperatures. As the temperature increases, the stability of Fe_3O_4 is favored over other corrosion products. Thus, for hot water and wet steam systems more oxygen is desirable [16] (Fig. 16). Here a critical concentration of oxygen, about 50 ppb is necessary to produce a tight protective magnetite film.

The benefit of a tenacious protective oxide film is also observed on Ti alloys [17] (Fig. 17). Thermally stabilizing the TiO_2 film on Ti at 700°C increases the resistance to erosion-corrosion at ambient temperature beyond that demonstrated by the natural TiO_2 film formed on pure Ti under ambient conditions. This difference between the naturally formed oxide and a high-temperature stabilized oxide can only be attributed to differences in film properties that are as yet undefined.

Temperature has a profound effect on the erosion-corrosion behavior of most metals. As expected, metal loss increases with escalating temperature. For instance, the oxide formed on 9 Cr–1 Mo steel at ambient temperature is either Cr_2O_3 or a Cr-substituted iron oxide,

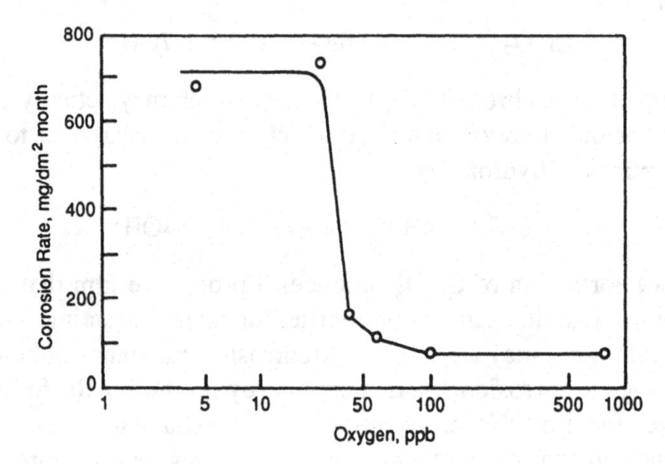

Figure 16. Corrosion rate dependence of carbon steel on oxygen content in water at 100°C.

depending on the environment. At higher temperatures Cr_2O_3 becomes the predominant corrosion film. However, under erosion-corrosion conditions, the composition and nature of the film are paramount to the resistance of the underlying metal. Figure 18 illustrates the response of 9 Cr–1 Mo to erosion-corrosion by air with alumina particles as a function of temperature [18]. The wastage was

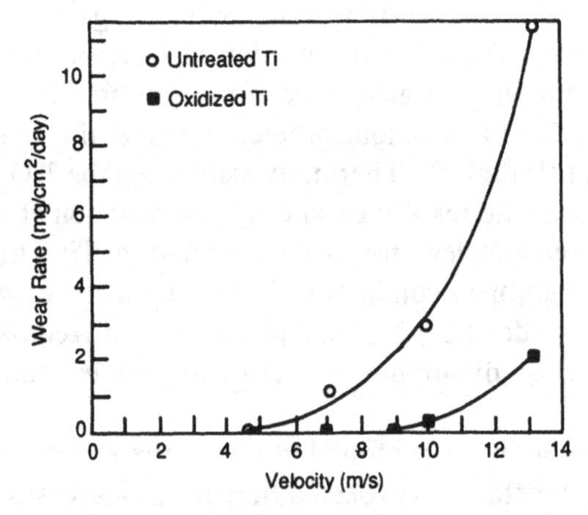

Figure 17. Erosion-corrosion behavior of untreated Ti and Ti oxidized in air at 700°C for 2 hr, exposed to aerated coal/water slurry.

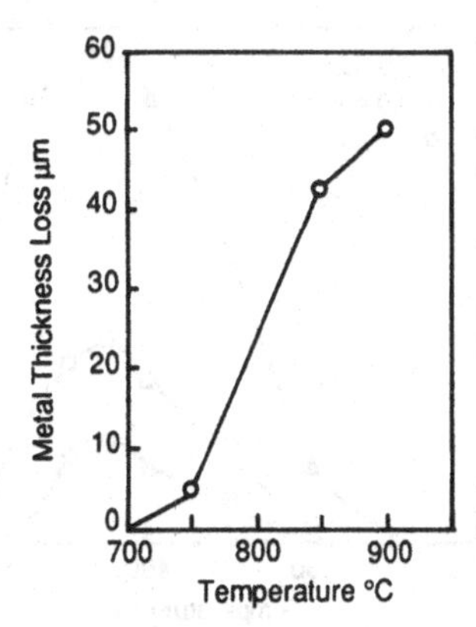

Figure 18. Erosion-corrosion of 9 Cr-1 Mo steel in air containing 130 Al_2O_3.

essentially negligible below 750°C due to the presence of an unidentified thin protective film. Between 750 and 850°C, the film composition changed, leading to a significantly higher degree of erosion-corrosion. However, above 850°C the film composition changes again, and the metal regains some erosion-corrosion resistance. The scale formed in the region of high erosion-corrosion rate was primarily α-Fe_2O_3. However, the film was patchy and did not produce a uniform coverage of the surface, thus allowing erosion and oxidation of the base metal independent of film properties. As the temperature increased to 850°C, a combined scale of α-Fe_2O_3 and the spinel $FeCrO_4$ formed. Above this temperature, a duplex scale of α-Fe_2O_3 and Cr_xO_y was present.

Of course, this transition behavior for erosion-corrosion as a function of scale composition and temperature is a function of the particular test environment. The presence or absence of solid particles versus a gas jet, the angularity of the particles, and the impingement angle will also affect this transition behavior. However, the conceptual ramifications of scale composition on erosion-corrosion are important here rather than the actual numbers.

Most recently it has been shown that when particle flux depends on velocity of the erodent, a maximum appears in the temperature dependence of erosion-corrosion for type 347 stainless steel (Fig. 19), Incoloy 800H, and carbon steel (Fig. 20) [19]. Regardless of particle

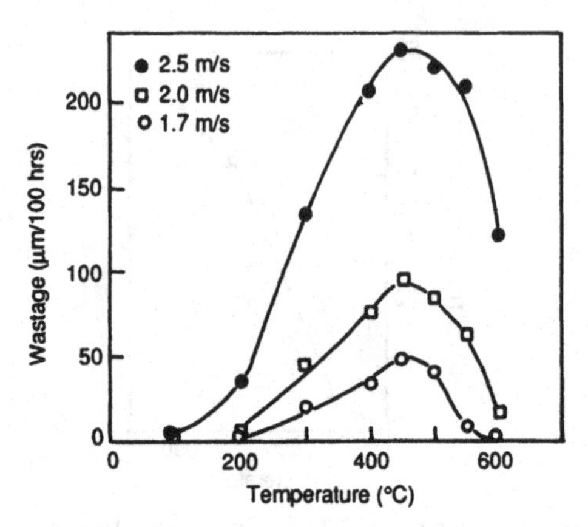

Figure 19. Erosion-corrosion of type 347 stainless steel as a function of velocity and temperature.

velocity, the highest erosion-corrosion for 347 was encountered at about 450°C, while for carbon steel it was approximately 300°C. The difference in maximum temperature between these two alloys is attributable to the film composition and properties. One explanation of this maximum for austentitic and ferritic alloys is the competition between film plasticity that increases with temperature versus oxidation rate, which also increases with temperature. Figure 21 illustrates the competing mechanisms [19]. Increasing ductility of the film with

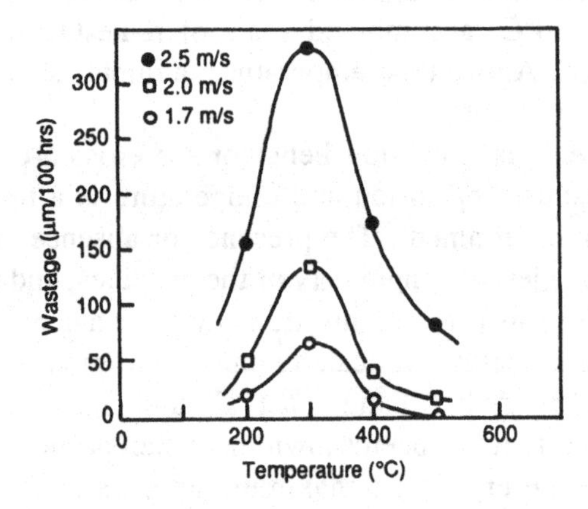

Figure 20. Erosion-corrosion of mild steel as a function of velocity and temperature.

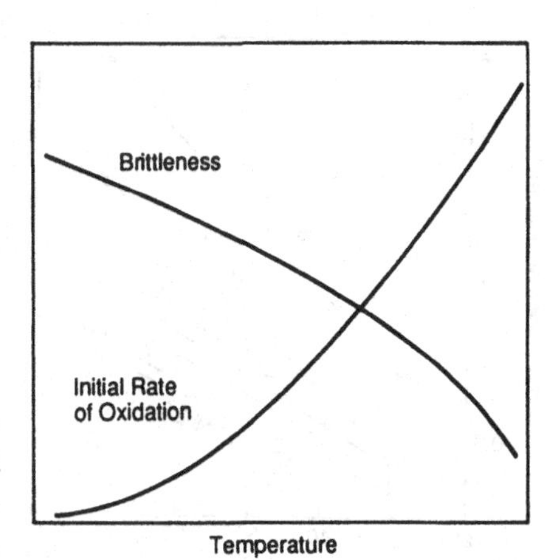

Figure 21. Schematic of the competition between oxidation of a metal and oxide plasticity (brittleness) as a function of temperature.

higher temperature provides greater resistance to impacting particles; however, when the film is spalled, the oxidation rate can be considerable. At low temperature, film brittleness is the controlling factor, whereas at high temperature oxidation becomes the controlling mechanism.

The presence of hard oxides of elements like Cr appear to enhance erosion-corrosion resistance. Therefore, it would be anticipated that the addition of small amounts of other elements that form ceramic or refractory oxides should increase erosion-corrosion resistance. Figure 22 demonstrates the combined effect of Cr and Si additives [20]. It is evident by comparing the 5 Cr, 0.5 Mo alloys that a relatively small addition of Si can have a major impact on the erosion-corrosion resistance of an alloy.

Generally, thicker films are found to be less erosion-corrosion resistant than thin films because of the additional stresses produced in the film itself during film growth. These stresses are then available during particle impact to aid spalling of large sections of the films. Thin films are often only chipped by the impinging particle and the film can easily be repaired. Also, this film will assist in transmitting the elastic or plastic strain of colliding particles to the base metal where thick films will accept most of this strain.

Thus fine-grain, thin ductile films are more advantageous than thick, large-grain films for erosion-corrosion resistance. In fact, the ductility of the film plays a major role in the resistance to erosion-

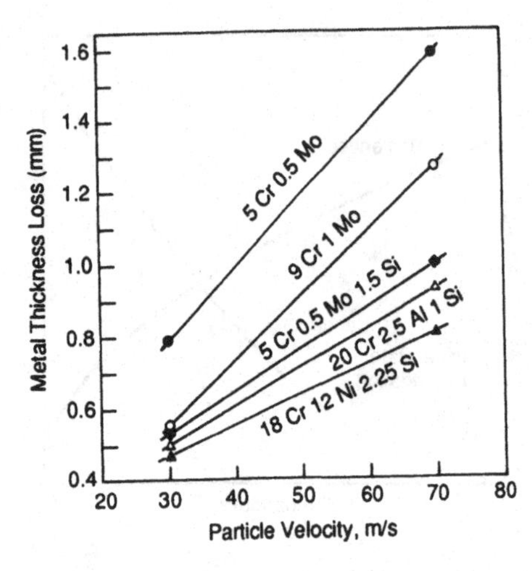

Figure 22. Effect of particle velocity and alloy composition on erosion-corrosion at 850°C in air. Particles were 130 μm Al_2O_3.

corrosion. Consider, for example, the oxide formed on nickel, which is NiO. There are no other stoichiometric forms of nickel oxide to introduce compositional variations in the picture [21]. A thermally grown oxide film, NiO, subjected to airflow with solid particles demonstrates this behavior in Fig. 23. Erosion-corrosion was greatest at ambient temperature compared to several higher temperatures, even though the tests at 650°C and 800°C were at about twice the erosion velocity as the ambient temperature tests. The appearance of the oxide surface after erosion-corrosion testing indicated the higher-temperature oxides were more resistant to impact by the impinging hard particles, thus suggesting a greater degree of plasticity (ductility) for the NiO film at high temperature compared to ambient temperature. However, since these tests were run with air, another plausible explanation is simply that oxidation kinetics are faster at higher temperatures. Therefore, film repair from each impact is faster and more comprehensive at temperatures higher than ambient. Regardless, it is evident from other considerations of particle energy and impact on surface that a ductile film is more desirable for enhanced erosion-corrosion resistance than is a brittle film.

Likewise, adhesion of the film to the metal substrate is crucial to erosion-corrosion resistance. Figure 24 compares the erosion-corrosion resistance of stainless steels and nickel-base alloys to erosion-corrosion resistance as a function of the alloy hardness at a test

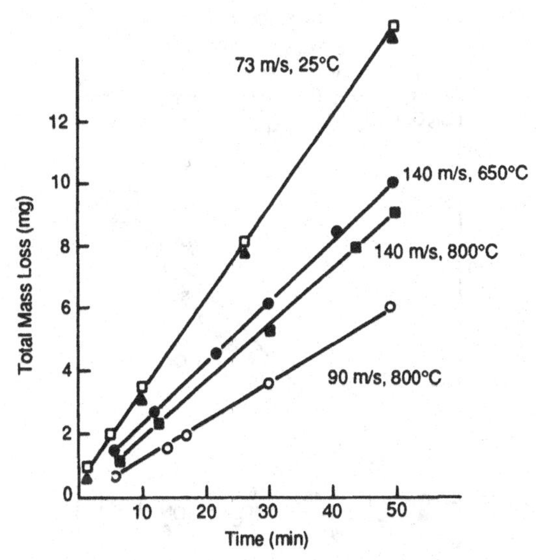

Figure 23. Erosion-corrosion behavior of preoxidized nickel as a function of velocity and temperature.

temperature of 760°C in a simulated flue gas environment [22]. While the data imply that increasing hardness of the metal reduces erosion-corrosion resistance, the more important parameter is film adhesion. Across the top of the figure the qualitative film adhesiveness is listed along with oxide composition. It is generally observed that either Al_2O_3 or Cr_2O_3 with excellent film adhesion provides greater erosion-corrosion resistance than alloys that produce a film of comparable composition but poor adhesion. Likewise, duplex or mixed scales may not be beneficial to erosion-corrosion due to a tendency for decohesion at the scale/scale interface.

Ideally, there are three regions of erosion-corrosion behavior with increasing erosion rate. In the first case, erosion of portions of the oxide will occur from the impacting particles. The second region will display a balance between the oxidation rate which is fast enough to match the erosion rate of the oxide; the net result is a loss of metal as the metal-oxide interface retreats into the metal. In the third region the erosion rate is so high that oxidation cannot keep up, and purely mechanical degradation of the metal itself occurs. In reality, these events are not separated by definitive boundaries but display a transi-

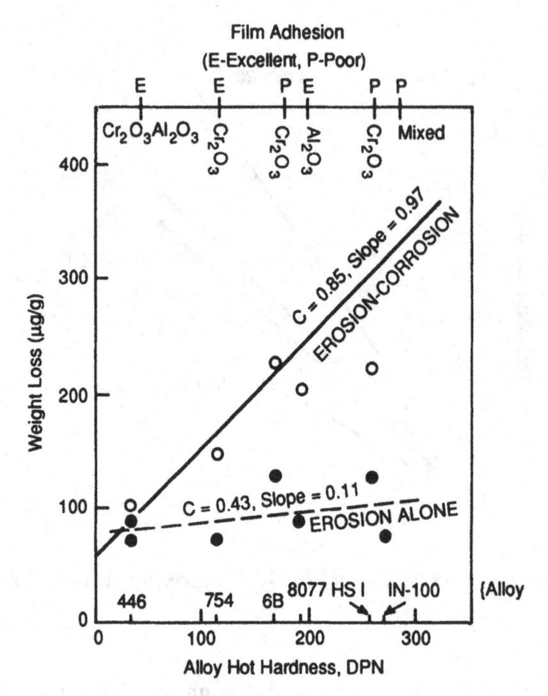

Figure 24. Erosion-corrosion behavior of stainless steel and nickel alloys as a function of metal hardness at 760°C.

tional behavior between these ideal regions. The specific regime is determined by a myriad of factors such as temperature, velocity, particle angularity, absence or presence of particles, particle composition, gas composition, impingement angle, etc.

Considerable work remains to be performed to understand the role of corrosion films in the erosion-corrosion behavior of metals and alloys. A good example of how other, similar investigations may be applied is illustrated in Fig. 25 [23]. These Mott-Schottky plots show the variation of lattice defects in ZnO as a function of surface treatment. Curve 1 was mechanically polished with diamond paste while curve 2 was etched for 5 sec in 5 *M* HCl. The dotted line represents behavior after photoillumination. Curve 1 represents a mechanically perturbed surface layer compared to curve 2, and, thus, corresponds to a higher donor concentration and a decreased capacitance after

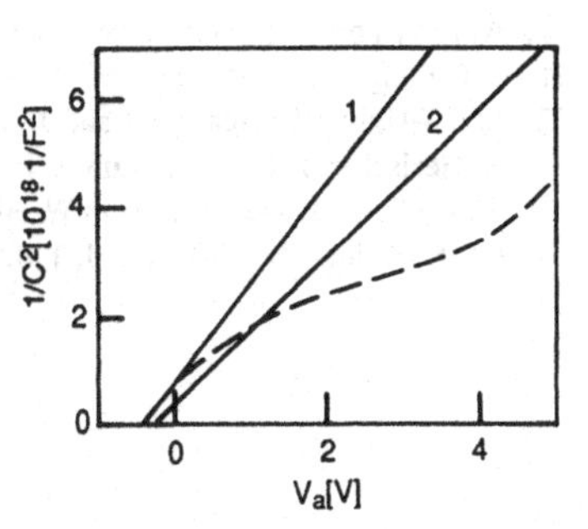

Figure 25. Mott-Schottky plots of $1/C^2$ versus electrode potential V_a (SCE) of [0001] ZnO single crystals (1) mechanically polished with diamond, (2) etch in 5 *M* HCl and dotted line, after illumination.

abrasion. Similar behavior should be expected under erosion-corrosion conditions.

Additional study of the relationship of defect type and concentration as a function of erosion should produce significant results in understanding this phenomenon.

References

1. U.R. Evans, *The Corrosion and Oxidation of Metals*, E. Arnold, London, 1960.
2. G. Engelberg and J. Yahalom, *Corrosion Sci.* **1972,** *12*, 469.
3. G.J. Danek, *Naval Engineers J.* **1966,** *78*, 763.
4. B.C. Syrett, *Corrosion* **1976,** *32*, 242.
5. K.D. Efird, *Corrosion* **1977,** *33*, 3.
6. M. Akkaya and J.R. Ambrose, *Corrosion* **1985,** *41*, 707.
7. J. Zahavi and H.J. Wagner, *Corrosion-Erosion Behavior of Materials*, p. 226, Ed. K. Natesan, TMS of AIME, Warrandale, PA, 1980.
8. Y. Watanabe, T. Shoji, and H. Takahashi, *Corrosion Eng.* (Japan) **1988,** *37*, 51.
9. H.G. Heitmann and W. Kastner, *VGB-Kraftwerkstechnik* **1982,** *63*, 180.
10. A. Ikeda, M. Ueda, and S. Mukai, Corrosion/83, Paper No. 45, Anaheim, CA, 1983.
11. M.G. Fontana and N.D. Greene, *Corrosion Engineering*, p. 75, McGraw-Hill, New York, 1967.
12. N.S. Hirota, *Corrosion*, vol. 13, p. 964, *Metals Handbook*, ASM International, 1987.
13. J. Postlethwaite, M.H. Dobbin, and K. Bergevin, *Corrosion* **1986,** *42*, 514.
14. J. Postlethwaite, *Mat. Perform.* **1987,** *26*, 41.
15. J. Postlethwaite, *Corrosion* **1981,** *37*, 1.
16. G.A. Delp, J.D. Robinson, and M.T. Sedlack, EPRI NP-3944, Electric Power Research Institute, April 1985.
17. G.R. Hoey and J.S. Bednar, *Corrosion* **1983,** *22*, 9.
18. A. Levy and Y-F. Man, *Wear* **1986,** *111*, 161.
19. A.J. Ninham, I.M. Hutchings, and J.A. Little, Corrosion/89, Paper No. 544, NACE, New Orleans, LA, 1989.

20. A. Levy and B-Q. Wang, Corrosion/88, Paper No. 147, St Louis, MO, 1988.
21. C.T. Kang, F.S. Pettit, and N. Birks, *Metall. Trans. A* **1987,** *18A,* 1785.
22. I.G. Wright, V. Nagarajan, and J. Stringer, *Oxidation of Metals* **1986,** *25,* 175.
23. H. Gerischer, F. Hein, M. Lubke, E. Meyer, B. Pettinger, and R. Schoppel, *Ber. Bunsenges Phys. Chem.* **1973,** *77,* 284. in W. Hirschwald, *Current Topics in Materials Science,* vol. 6, Ed. E. Kaldis, North-Holland, 1980.

Index